Lecture Notes in Computer Science 10480

Commenced Publication in 1973
Founding and Former Series Editors:
Gerhard Goos, Juris Hartmanis, and Jan van Leeuwen

Ngoc Thanh Nguyen · Ryszard Kowalczyk
Jacek Mercik (Eds.)

Transactions on Computational Collective Intelligence XXVII

Editors-in-Chief

Ngoc Thanh Nguyen
Wrocław University of Science
and Technology
Wrocław
Poland

Ryszard Kowalczyk
Swinburne University of Technology
Hawthorn, VIC
Australia

Guest Editor

Jacek Mercik
Department of Finance and Management
The WSB University
Wrocław
Poland

ISSN 0302-9743   ISSN 1611-3349   (electronic)
Lecture Notes in Computer Science
ISSN 2190-9288   ISSN 2511-6053   (electronic)
Transactions on Computational Collective Intelligence
ISBN 978-3-319-70646-7   ISBN 978-3-319-70647-4   (eBook)
https://doi.org/10.1007/978-3-319-70647-4

Library of Congress Control Number: 2017958562

This Springer imprint is published by Springer Nature
The registered company is Springer International Publishing AG
The registered company address is: Gewerbestrasse 11, 6330 Cham, Switzerland

Transactions on Computational Collective Intelligence XXVII

Preface

It is my pleasure to present to you the XXVII volume of LNCS *Transactions on Computational Collective Intelligence*. In Autumn 2016 (November 25) at the WSB University in Wroclaw, Poland, there was the second seminar on "Quantitative Methods of Group Decision-Making." Thanks to WSB University in Wroclaw, we had an excellent opportunity to organize and financially support the seminar. This volume presents post-seminar papers of participants to this seminar. During the seminar, we listened to and discussed over 18 presentations from 15 universities. The XXVII issue of TCCI contains 13 high-quality, carefully reviewed papers.

The first paper "Kalai–Smorodinsky Balances for n-Tuples of Interfering Elements" by David Carfi, Alessia Donato, and Gianfranco Gambarelli is devoted[1] to studying a mathematical game model providing optimal Kalai–Smorodinsky compromise solution n-tuples, whose components indicate active principle quantities, in a specific non-linear interfering scenario, with n possible interacting elements. The problem was solved by using the Carfi's pay-off analysis method for differentiable pay-off functions and implementing the Matlab algorithms for the construction and representation of the pay-off spaces and for the finding of Kalai–Smorodinsky solutions. The core section of the paper studies the game in the n-dimensional case, by finding the critical zone of the game in its Cartesian form together with proof of a theorem and a lemma about the Jacobian determinant of the n-game. In a particular highly symmetrical case, the analytical solution of the Kalai–Smorodinsky compromise problem is presented too.

In the second paper entitled "Reason vs. Rationality: From Rankings to Tournaments in Individual Choice" by Janusz Kacprzyk, Hannu Nurmi, and Slawomir Zadrozny, one may find the standard assumption in decision theory, microeconomics, and social choice that individuals (consumers, voters) are endowed with preferences that can be expressed as complete and transitive binary relations over alternatives (bundles of goods, policies, candidates). While this may often be the case, the authors show by way of toy examples that incomplete and intransitive preference relations are not only conceivable, but make intuitive sense and they suggest that fuzzy preference relations and solution concepts based on them are plausible in accommodating those features that give rise to intransitive and incomplete preferences.

In the third paper, "A Note on Positions and Power of Players in Multicameral Voting Games," Marcin Malawski presents a study of a multicameral simple game as an intersection of a number of simple games played by the same set of players: A coalition is winning in the multicameral game if and only if it is winning in all the individual games played. Examples include decision rules in multicameral parliaments where a bill must be passed in all the houses of the parliament, and voting rules in the European

---

[1] Hereafter description of the papers are directly taken from summaries prepared by their authors.

Union Council where a winning coalition of countries must satisfy two or three independent criteria. The presented paper is a preliminary study of relations between the positions and power indices of players in the "chamber" games and in the multicameral game obtained as the intersection. The author demonstrates that for any power index satisfying a number of standard properties, the index of a player in the multicameral game can be smaller (or greater) than in all the chamber games; this can occur even when the players are ordered the same way by desirability relations in all the chamber games. He also observes some counterintuitive effects when comparing the positions and decisiveness of players. However, as expected, introducing an additional chamber with all the players equal (a one man–one vote majority game) to a complete simple game reduces all the differences between the Shapley–Shubik indices of players.

The fourth paper "On Ordering a Set of Degressively Proportional Apportionments" by Katarzyna Cegielka, Piotr Dniestrzanski, Janusz Lyko, Arkadiusz Maciuk, and Radoslaw Rudek proposes a solution to the most important problem in a practical implementation of degressive proportionality, namely, its ambiguity. They introduce an order relation on a set of degressively proportional allocations. Its main idea is to define greater allocations such that emerge from others after transferring a certain quantity of goods from smaller to greater entities contending in distribution. Thus, maximal elements in this ordering are indicated as the sought-after solution sanctioning boundary conditions as the only reason for moving away from the fundamental principle of proportionality. In case of several maximal elements, the choice of one allocation remains an open issue, but the cardinality of the set from which they make a choice can be reduced significantly. In the best-known example of application of degressive proportionality, which is the apportionment of seats in the European Parliament, the considered set contains a maximal element. Thereby, there exits an allocation that is nearest to the proportional distribution with respect to transfer relation.

In the fifth paper entitled "Preorders in Simple Games," Josep Freixas and Montserrat Pons consider a hierarchy among players in a simple game with total preorder given by any power index. The desirability relation, which is also a preorder, induces the same hierarchy as the Banzhaf and the Shapley indices on linear games, i.e., games in which the desirability relation is total. The desirability relation is a sub-preorder of another preorder, the weak desirability relation, and the class of weakly linear games, i.e., games for which the weak desirability relation is total, is larger than the class of linear games. The weak desirability relation induces the same hierarchy as the Banzhaf and the Shapley indices on weakly linear games. They define a chain of preorders between the desirability and the weak desirability preorders. From them they obtain new classes of totally preordered games between linear and weakly linear games.

In the sixth paper "Sub-coalitional Approach to Values," Izabella Stach analyzes the behavioral models of classic values (like the Shapley and Banzhaf values) by considering the contributions to coalition S as contributions delivered by the players individually joining such a coalition as it is being formed; i.e., $v(S) - v(S\setminus\{i\})$. In this paper, she proposes another approach to values where these contributions are considered as given by sets of players: $(v(S) - v(S\setminus R))$, where S, R are subsets of the set of all players involved in cooperative game v. Based on this new approach, several sub-coalitional values are proposed, and some properties of these values are shown.

In the seventh paper entitled "The Effect of Brexit on the Balance of Power in the European Union Council: An Approach Based on Precoalitions," Jacek Mercik and David M. Ramsey investigate the change in the balance of power in the European Union Council due to the United Kingdom leaving. This analysis is based on the concept of power indices in voting games where natural coalitions, called precoalitions, may occur between various players (or parties). The precoalitions in these games are assumed to be formed around the six largest member states (after Brexit, the five largest), where each of the remaining member states joins the precoalition based around the large member state which is the most similar according to the subject of the vote. They consider adaptations of three classic indices: the Shapley–Shubik, Banzhaf–Penrose, and Johnston indices based on the concept of a consistent share function (also called quotient index). This approach can be interpreted as a two-level process of distributing power. At the upper level, power is distributed among precoalitions. At the lower level, power is distributed amongst the members of each precoalition. One of the conclusions of the research is that removing the UK from the voting game means that the power indices of small countries actually decrease. This seems somewhat surprising as the voting procedure in the EU council was designed to be robust to changes in the number and size of member states. This conclusion does not correspond to a general result, but does indicate the difficulty of defining voting rules which are robust to changes in the set of players.

The eighth paper entitled "Comparison of Voting Methods Used in Some Classical Music Competitions" by Honorata Sosnowska is devoted to a comparison of the rules of voting in the last two main Polish classical music competitions: the XVIIth Chopin Piano Competition and the XVth Wieniawski Violin Competition. Weak and strong points of rules are analyzed. The rules are also compared with rules used in the previous editions of the competitions. The author concludes that the changes resulted in the simplification of rules.

In the ninth paper "Determinants of the Perception of Opportunity" by Aleksandra Sus, the determinants of the perception of opportunity are analyzed. Contemporary strategic management has accepted the category of opportunity, although it cannot be reflected in the organization's plans and strategies. Alertness, proactivity, social networks, and knowledge resources are the categories that come up most often when discussing opportunity perception as one of the determinants of entrepreneurial activity. In reality, they are the result of both behavioral and cognitive processes. The purpose of the article is to identify the primary factors that predetermine the idiosyncrasy of how opportunity is perceived by various persons, such as creativity, intuition, and divergent thinking. The article presents opportunity value chains. The article also discusses the process of group decision-making in terms of opportunity.

The tenth paper entitled "Free-Riding in Common Facility Sharing" is authored by Federica Briata and Vito Fragnelli. The paper deals with the free-riding situations that may arise from sharing maintenance costs of a facility among its potential users. The non-users may ask for a check to assess who the users are, but they have to pay the related cost; consequently, a non-user may not ask for the check, with the hope that the other non-users ask and pay for it. In this paper, they provide incentives for asking for the check, without suffering a higher cost.

The 11th paper is the joint work of Natalie van der Wal, Daniel Formolo, Mark A. Robinson, Michael Minkov, and Tibor Bosse. The paper is entitled "Simulating Crowd

Evacuation with Sociocultural, Cognitive, and Emotional Elements." In this research, the effects of culture, cognitions, and emotions on crisis management and prevention are analyzed. An agent-based crowd evacuation simulation model was created, named IMPACT, to study the evacuation process from a transport hub. To extend previous research, various sociocultural, cognitive, and emotional factors were modeled, including: language, gender, familiarity with the environment, emotional contagion, prosocial behavior, falls, group decision-making, and compliance. The IMPACT model was validated against data from an evacuation drill using the existing EXODUS evacuation model. Results show that on all measures, the IMPACT model is within or close to the prescribed boundaries, thereby establishing its validity. Structured simulations with the validated model revealed important findings, including: the effect of doors as bottlenecks, social contagion speeding up evacuation time, falling behavior not affecting evacuation time significantly, and traveling in groups being more beneficial for evacuation time than traveling alone.

The 12th paper "Group Approximation of Task Duration and Time Buffers in Scrum" is written by Barbara Gładysz and Andrzej Pawlicki. Expansion of modern IT technologies, which took place in the past few years, caused a significant increase in software projects. These projects are quite often complex ventures burdened with high risk. Nowadays, a large number of software projects is managed using the Scrum framework. In Scrum, where people form self-organizing team, group decisions became an essential element of the project, which plays an important role in creating time approximation or in managing potential risks. This paper focuses on group decisions, temporal aspects of estimation, and risk management in the Scrum project. In the article they present a conceptual model of the extension of the Scrum framework by risk management processes and project time estimation. The proposed model contains time buffers based on mixture probability distribution, which improve the Scrum framework in terms of group estimation.

Traditionally, the last paper is an invited paper, and in this volume it is entitled "Extending Estimation of Distribution Algorithms with Agent-based Computing Inspirations" authored by Aleksander Byrski, Marek Kisiel-Dorohinicki, and Norbert Tusinski. In their paper, several extensions of a successful EDA-type algorithm, namely, COMMAop, inspired by the paradigm of agent-based computing (EMAS) are presented. The proposed algorithms leveraging notions connected with EMAS, such as reproduction and death, or even the population decomposition, turn out to be better than the original algorithm. The evidence for this is presented at the end of the paper, utilizing QAP problems by Eric Taillard as benchmarks.

I would like to thank all the authors for their valuable contributions to this issue and all reviewers for their feedback, which helped to keep the papers of high quality. My very special thanks go to Prof. Ngoc-Thanh Nguyen, who encouraged us to prepare this volume, and to Dr. Bernadetta Maleszka, who helped us publish this issues in due time and in good order.

May 2017   Jacek Mercik

Transactions on Computational Collective Intelligence

This Springer journal focuses on research in applications of the computer-based methods of computational collective intelligence (CCI) and their applications in a wide range of fields such as the Semantic Web, social networks, and multi-agent systems. It aims to provide a forum for the presentation of scientific research and technological achievements accomplished by the international community.

The topics addressed by this journal include all solutions to real-life problems for which it is necessary to use computational collective intelligence technologies to achieve effective results. The emphasis of the papers published is on novel and original research and technological advancements. Special features on specific topics are welcome.

Editor-in-Chief

Ngoc Thanh Nguyen   Wrocław University of Science and Technology, Poland

Co-Editor-in-Chief

Ryszard Kowalczyk   Swinburne University of Technology, Australia

Editorial Board

John Breslin	National University of Ireland, Galway, Ireland
Longbing Cao	University of Technology Sydney, Australia
Shi-Kuo Chang	University of Pittsburgh, USA
Oscar Cordon	European Centre for Soft Computing, Spain
Tzung-Pei Hong	National University of Kaohsiung, Taiwan
Gordan Jezic	University of Zagreb, Croatia
Piotr Jędrzejowicz	Gdynia Maritime University, Poland
Kang-Huyn Jo	University of Ulsan, Korea
Yiannis Kompatsiaris	Centre for Research and Technology Hellas, Greece
Jozef Korbicz	University of Zielona Gora, Poland
Hoai An Le Thi	Lorraine University, France
Pierre Lévy	University of Ottawa, Canada
Tokuro Matsuo	Yamagata University, Japan
Kazumi Nakamatsu	University of Hyogo, Japan
Toyoaki Nishida	Kyoto University, Japan
Manuel Núñez	Universidad Complutense de Madrid, Spain
Julian Padget	University of Bath, UK
Witold Pedrycz	University of Alberta, Canada

Contents

Kalai-Smorodinsky Balances for N-Tuples of Interfering Elements

David Carfi[1], Alessia Donato[2(✉)], and Gianfranco Gambarelli[3]

[1] University of California at Riverside, UCRiverside, Riverside, USA
`dcarfi@unime.it, davidcarfi@gmail.com`
[2] University of Messina, Messina, Italy
`donatoalessia89@gmail.com`
[3] University of Bergamo, Bergamo, Italy
`gianfranco.gambarelli@unibg.it`

Abstract. The study proposed here builds up a game model (with associated algorithms) in a specific non-linear interfering scenario, with n possible interacting elements. Our examination provides optimal Kalai-Smorodinsky compromise solution n-tuples, for the game, whose components indicate active principle quantity percentages. We solve the problem by using the Carfi's payoff analysis method for differentiable payoff functions. Moreover we implement Matlab algorithms for the construction and representation of the payoff spaces and for the finding of Kalai-Smorodinsky solutions. The software for the determination of graphs are adopted, but not presented here explicitly. The core section of the paper, completely studies the game in the n-dimensional case, by finding the critical zone of the game in its Cartesian form. At this aim, we need to prove a theorem and a lemma about the Jacobian determinant of the n-game. In the same section, we write down the intersection of the critical zone and the Kalai-Smorodinsky straight-line. In the Appendix 1 we solve the problem in closed form for the 2 dimensional case and numerically for $n > 2$. Our methods works also for games with non-convex payoff space. Finally, in a particular highly symmetrical case, we solve analytically the Kalai-Smorodinsky compromise problem in all cases. We provide some applications of the obtained results, particularly to economic problems.

Keywords: Antagonistic elements · Interfering elements · Optimal dosage · Synergies · Complete study of differentiable game · Kalai-Smorodinsky solutions

1 Introduction

The paper deals with a balance problem for n interfering elements, proposing a version of Kalai-Smorodinsky solution, by modeling the interaction (interference) using a particular class of normal form games in which each player dispose of two pure strategies. Moreover, for each player one strategy strictly dominates the

© Springer International Publishing AG 2017
N.T. Nguyen et al. (Eds.): TCCI XXVII, LNCS 10480, pp. 1–27, 2017.
https://doi.org/10.1007/978-3-319-70647-4_1

other, so the game necessarily has a unique Nash equilibrium. When this equilibrium is not Pareto optimal (leading to an applicative significant case) it makes sense to select some strictly better point from the Pareto frontier. One plausible solution is that offered by the Kalai-Smorodinsky method as presented in the paper, with the equilibrium constituting the "disagreement point". Since this solution results necessarily from playing mixed strategies, finding these strategies is also of some interest from the mathematical point of view, not only from an economic and applicative point of view. An analytic method for detecting them by finding critical points of the payoff (vector) function is presented, together with explicit formulae for the Kalai-Smorodinsky solution derived from it.

1.1   Structure of the Paper

The paper is organized as follows.

- In Sect. 2, we present the model.
- In Sect. 3, we develop our proposed model (by using the Carfi's payoff analysis method for differentiable payoff functions) in dimension 2, studying the game in the finite strategy case and in the infinite strategy case. In particular, we consider elements interfering positively and negatively. Then we solve the problem in closed analytical form for

$$a + b < 1,$$

  by a Kalai-Smorodinsky type compromise solution.
- In Sect. 4, the core section of the paper, we completely study the game in the n-dimensional case, by finding the critical zone of the game in its Cartesian form. We write down the intersection of the critical zone and the Kalai-Smorodinsky straight-line and we solve numerically the problem, for

$$a + b < 1.$$

  In a particular highly symmetrical case, we solve analytically the Kalai-Smorodinsky compromise problem.
- In Sects. 5, 6 and 7, we discuss possible applications of the proposed model and we present the conclusions and the conclusive remarks of the paper.
- In Appendix 1, algorithms for determining solutions are provided, together with softwares and examples for the case $n = 2$, $n = 10$.
- In Appendix 2, we present the necessary preliminaries and notations about game theory which are not supposed to be known to the reader.

2   The Model

In this paper, we consider a model of strategic interaction for n different factors (elements), which we consider as interfering elements (i.e. drugs, anticryptogamics, investments, commodities and so on) and related effects resulting from their use (e.g., curing diseases, killing parasites, gains, commodity demand and so on), following a precise law of interaction. This law of interaction states that if no elements are used, then all the effects are null and if a single element is employed in the optimal using dosage alone, then the level of the relative effect is 1, while the level of the effect for the others is null. The general model, we consider here, is represented by an n-player differentiable general sum game. Indicating by $q \in [0,1]^n$ the n-percentage profile defined as follows $q = (q_1, q_2, \ldots, q_n)$, we can write the payoff function of the game by the n-vector

$$f(q) = \left(q_j - (1 - a_j) \prod_{i=1}^{n} q_i \right)_{j=1}^{n},$$

for every q belonging to the hypercube $[0,1]^n$.

2.1 Applicative Motivations

Often decisions in different situations and contexts need to consider the joined effects of various elements that might interfere with each other. For example, in Industrial Economics the demand of an asset may be influenced by the supply of other assets, with synergic or antagonistic effects or with a convex combination of the two extreme scenarios. The same situation might happen in Public Economics, where different economic policies may create mutual interference. Analogously, in Medicine with drugs whose combined administration might produce extra damages or synergies. Other examples occur in Agriculture, Zootechnics and so on. When it appears necessary to intervene in such situations, there exists sometimes a primary interest for one effect rather than others.

A previous model by Carfi et al. [9], calculates the quantities of two elements that interfere with each other, optimizing the required ratio of the effects and taking into account the minimum quantities to be allocated. That former model can be applicable to the case of two elements smoothly interfering, or better, interfering in a differentiable fashion. Indeed, the main theorem we use in that paper in order to construct the analytic game theory model requires and adopts differentiability assumptions. Such a differentiable methods allows to obtain, exactly, the compromise solution vector in a best compromise setting within a differentiable but not necessarily convex decision problem. Nevertheless, the extension to the case of payoff functions which are discontinuous or not differentiable on a finite subset of a strategy space, appears possible with some technicalities.

Here we go deeply inside the previous two dimensional case in order to clarify our techniques and some aspects and to prepare the field to the n-dimensional case, that constitutes the core of this paper. The compromise solution, that is the solution of our decision problem, are determined, fixed the dimension of the problem, by an original algorithm which can be applied with the chosen computational precision. The algorithm was written with the applicative software Matlab, obviously based on the analytical methods proposed by Carfi in [7]. In the algorithm we concentrate upon the determination of a Kalai-Smorodinsky type compromise solution.

3   The Mathematical Model: 2-Dimensional Case

To better understand the meaning of our model, we analyze - for simplicity - the two-dimensional case. We distinguish the finite case and the infinite case. In the finite case the game is represented by a bi-matrix

$$M = \begin{pmatrix} A' & D' \\ B' & C' \end{pmatrix} = \begin{pmatrix} (0,0) & (0,1) \\ (1,0) & (a,b) \end{pmatrix},$$

whose elements correspond to the effects of the profile strategies A, B, C, D shown in Fig. 1. Every profile strategy indicates a pair of quantity percentages of the two elements respectively used. In the finite case, the possible values of each element are 0 or 1 (0 means that we don't use the corresponding element and 1 means that we use 100% of the element required quantity for its scope).

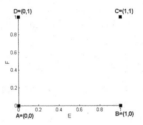

Fig. 1. Bi-strategy space of the finite game M.

In particular, the profile strategy $C = (1,1)$, representing the use of a percentage quantity 100% for both elements, appears important because it is the unique Nash equilibrium of the game. The effect of the Nash equilibrium is, by definition, a couple (a, b). The effects a and b are positive real numbers. Moreover, we observe that:

- if $a, b < 1$, then the interaction can be considered antagonistic;
- if $a, b > 1$, then the interaction can be considered synergic;
- if $a, b = 1$, then the interaction can be considered neutral;
- if $a > 1$ and $b \leq 1$, the interaction helps the action of the first element while depressing (or leaving unchanged) the action of the second element.

The payoff space of that finite game, according to the value of $a + b$, can be distinguished in three cases (see Fig. 2).
In the infinite game the bi-strategy space is represented by the square

$$E \times F = [0, 1] \times [0, 1],$$

see Fig. 3.

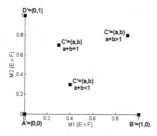

Fig. 2. Payoff space of the finite game with $a + b \lesseqgtr 1$.

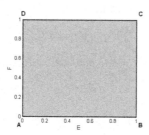

Fig. 3. Bi-strategy space for infinite game f.

3.1 Payoff Functions

Our payoff function in the general case will be the mixed extension of the finite game

$$f : E \times F \to \mathbb{R}^2,$$

expressing the levels of the effects, when the bi-percentage profile (p, q) is employed, with p chosen in $E = [0, 1]$ and q chosen in $F = [0, 1]$. The effect function (payoff function) is indeed defined on the bi-percentage space $E \times F = [0, 1]^2$, by the two components

$$
\begin{aligned}
f_1(p, q) &= p(1 - q) + apq = p - (1 - a)pq, \\
f_2(p, q) &= q(1 - p) + bpq = q - (1 - b)pq,
\end{aligned}
$$

for every percentage profile (p, q). The above very specific payoff function of the game is so defined by

$$f(p, q) = (p - a'pq, q - b'pq) = (p(1 - a'q), q(1 - b'p)), \tag{1}$$

where

$$a' = 1 - a, \quad b' = 1 - b$$

are the complements with respect to 1 of a and b, respectively.

3.2   Critical Zone of the Game

The Jacobian matrix of the function f at the bistrategy (p, q) is

$$J_f(p, q) = \begin{pmatrix} 1 - (1-a)q & -(1-a)p \\ -(1-b)q & 1-(1-b)p \end{pmatrix} = \begin{pmatrix} 1 - a'q & -a'p \\ -b'q & 1 - b'p \end{pmatrix}.$$

The Jacobian determinant at (p, q) is

$$\det J_f(p, q) = (1 - a'q)(1 - b'p) - a'b'pq$$
$$= 1 - a'q - b'p.$$

It vanishes upon the line r of equation

$$a'q + b'p = 1. \tag{2}$$

This line r intersects the bi-strategic space $E \times F = [0, 1]^2$ iff

$$a/b' \le 1$$

that is, iff

$$a + b \le 1.$$

In particular, that line is reduced to a point C for

$$a + b = 1.$$

The end points of this segment are the points

$$H = (a/b', 1) \quad K = (1, b/a')$$

(note that the relation $a/b' \le 1$ is equivalent to the relation $b/a' \le 1$). Consequently, the critical zone of the game is the segment $[H, K]$, and its first and second projections are, respectively, the interval

$$[a/b', 1]$$

and the interval

$$[b/a', 1]$$

(see for example in Fig. 4 a bi-strategy space of the game f for fixed values of a and b). If $a/b' > 1$, that is

$$a + b > 1,$$

the critical zone of the game is void.

We show in Figs. 5, 6 and 7 the payoff spaces in the three cases, for fixed values of a and b. Recalling that, for

$$a + b < 1$$

we need to transform the critical zone together with the sides of bi-strategic square, while in other cases we need to transform only the four sides of bi-strategic square, because the critical zone is void or a single point belonging to the boundary of the square.

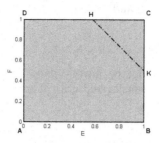

Fig. 4. Bi-strategy space of game f with $a = 0.4$ and $b = 0.3$.

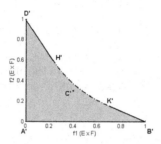

Fig. 5. Payoff space of game f with $a = 0.4$ and $b = 0.3$.

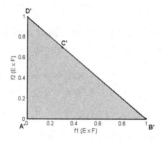

Fig. 6. Payoff space of game f with $a = 0.3$ and $b = 0.7$.

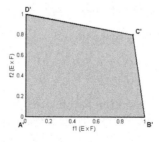

Fig. 7. Payoff space of game f with $a = 0.9$ and $b = 0.8$.

As we can observe in Figs. 5, 6 and 7:

- for
$$a + b < 1,$$

the point C' remains inside the payoff space, so we can search for a best compromise point in the transformed critical zone (see Fig. 5);
- for
$$a + b = 1,$$

although we find infinitely many indifferent Pareto points, constituting the diagonal $[D', B']$, we find a unique compromise point C' (see Fig. 6);
- for
$$a + b > 1,$$

the compromise solution is again the payoff C', corresponding to the Nash equilibrium (see Fig. 7).

We note that our best compromise derives from the application of a classic Kalai-Smorodinsky method using the Nash payoff as a threat point of the decision problem.

3.3   Kalai-Smorodinsky Solution

As we noted above, we need to find a compromise solution for the case
$$a + b < 1.$$

We decide to use the Kalai-Smorodinsky method and seek the solution directly in the bi-strategy space $E \times F$. The payoff functions are

$$f_1(p, q) = p(1 - q) + apq = p - (1 - a)pq,$$
$$f_2(p, q) = q(1 - p) + bpq = q - (1 - b)pq,$$
(3)

for every percentage profile (p, q).

We consider, in the payoff space $f(E \times F)$, the line passing through the Nash point $C' = (a, b)$ and the supremum point $(1, 1)$ of the payoff space. In parametric form, using (X, Y) as coordinates for the payoff universe, we can write:

$$(X, Y) = (a, b) + t(1 - a, 1 - b),$$

with t belonging to the real line, that is

$$\begin{cases} X = a + t(1 - a), \\ Y = b + t(1 - b). \end{cases}$$
(4)

We f-transfer the Eq. 4 (regardind the payoff space) in the bi-strategy space, by replacing on to the coordinates (X, Y) the corresponding expressions 3. So we get:

$$\begin{cases} f_1(p, q) = p - (1 - a)pq = a + t(1 - a), \\ f_2(p, q) = q - (1 - b)pq = b + t(1 - b). \end{cases}$$

From the first above equation we obtain

$$pq = \frac{p-a}{1-a} - t,$$

and then, by using the second equation, we obtain

$$q = b + \frac{1-b}{1-a}(p-a), \tag{5}$$

that is the Cartesian equation of the Kalai-Smorodinsky straight-line in the by-strategy space. Now, we find - in the bi-strategy space $E \times F$ - the intersection point between the *critical zone* of the game 2 and the straight line 5

$$\begin{cases} q(a-1) + p(b-1) + 1 = 0 \\ q = b + \dfrac{1-b}{1-a}(p-a). \end{cases}$$

From the system we obtain the solution with coordinates (see Appendix 1)

$$p^* = -\frac{a-b+1}{2b-2},$$

$$q^* = -\frac{b-a+1}{2a-2}.$$

That intersection point is the solution of the game by Kalai-Smorodinsky method. The point $S = (p^*, q^*)$ belongs to the space $E \times F$, when the parameter pair (a, b) varies in the open triangle with vertices $(0,0)$, $(1,0)$ and $(0,1)$.

Remark 1. We observe that the above Kalai-Smorodinsky straightline

$$(X, Y) = (a, b) + t(1 - a, 1 - b),$$

in the payoff space appears transformed by the inverse relation f^{-1}, of the payoff vector function, on to the straightline

$$q = b + \frac{1-b}{1-a}(p-a),$$

contained in the bi-strategy space.

3.4 Numerical Sample

For $a = 0.05$ and $b = 0.1$, we obtain (see Fig. 8):

$$p^* = 0.5278,$$

$$q^* = 0.5526.$$

In payoff space the effects are represented by (see Fig. 9):

$$f_1(p^*, q^*) = p^* - (1-a)p^*q^* \quad \Rightarrow \quad f_1(p^*, q^*) = 0.2507,$$

$$f_2(p^*, q^*) = q^* - (1-b)p^*q^* \quad \Rightarrow \quad f_2(p^*, q^*) = 0.2901.$$

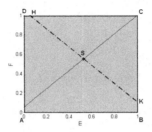

Fig. 8. Kalai-Smorodinsky solution in bi-strategy space for $a = 0.05$ and $b = 0.1$.

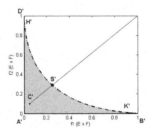

Fig. 9. Kalai-Smorodinsky solution in payoff space for $a = 0.05$ and $b = 0.1$.

4 n-Dimensional Study

By means of the Kalai-Smorodinsky method we can study the general game with n interfering elements.

4.1   Payoff Functions of the n-dimensional Game

Indicating by $q \in \mathbb{R}^n$ the n-percentage profile defined as follows

$$q = (q_1, q_2, \ldots, q_n),$$

we can write the payoff functions, in the n-dimensional case, in the following fashion:

$$f_1(q) = q_1 - (1 - a_1) \prod_{i=1}^n q_i,$$
$$f_2(q) = q_2 - (1 - a_2) \prod_{i=1}^n q_i,$$
$$\vdots$$
$$f_n(q) = q_n - (1 - a_n) \prod_{i=1}^n q_i,$$

for every q belonging to the hypercube E^n. So that, the payoff function of the game is compactly defined by n-vector

$$f(q) = \left(q_j - (1 - a_j) \prod_{i=1}^n q_i \right)_{j=1}^n, \tag{6}$$

for every q belonging to the hypercube E^n.

4.2   Jacobian Matrix of the Game

In order to find the critical zone of the game $(f, >)$, we need to calculate the Jacobian matrix of the function f at any point q, that is the matrix

$$
J_f(q) = \begin{pmatrix} \partial_1 f_1(q) & \cdots & \partial_n f_1(q) \\ \partial_1 f_2(q) & \cdots & \partial_n f_2(q) \\ \vdots & \vdots & \vdots \\ \partial_1 f_n(q) & \cdots & \partial_n f_n(q) \end{pmatrix},
$$

for every q belonging to the strategy space E^n. Let us calculate the derivative $\partial_j f_j$, for every positive integer $j \le n$. We easily obtain

$$
\partial_j f_j(q) = 1 + \left(\prod_{i=1, i \ne j}^{n} q_i \right) (a_j - 1),
$$

for every q belonging to the strategy space E^n and for every positive integer $j \le n$. Let us calculate the derivative $\partial_k f_j$, with $k \ne j$. We obtain

$$
\partial_k f_j(q) = \left(\prod_{i=1, i \ne k}^{n} q_i \right) (a_j - 1),
$$

for every q belonging to the strategy space E^n and for every positive integers $j, k \le n$, with $k \ne j$. In general, the derivative $\partial_k f_j$, is

$$
\partial_k f_j(q) = \delta_{kj} + \left(\prod_{i=1, i \ne k}^{n} q_i \right) (a_j - 1),
$$

for every q belonging to the strategy space E^n and for every positive integers $j, k \le n$, where δ represents the Kronecker delta.

More conveniently, we consider the following n-vectors:

– firstly, the vector

$$
a' := (a_i')_{i=1}^{n},
$$

where

$$
a_i' := 1 - a_i
$$

for every positive integer $i \le n$;
– secondly, the vector

$$
v := \left(\prod_{i=1, i \ne j}^{n} q_i \right)_{j=1}^{n},
$$

in other terms, the component v_j of the vector v is defined as the product

$$
\prod_{i=1, i \ne j}^{n} q_i,
$$

for every positive integer $j \le n$.

With the above convenient notations, we can finally write down the expression of the Jacobian matrix as follows:

$$J_f(q) = \left(\partial_k f_j(q) \right)^n_{j,k=1} = \left(\delta_{kj} - v_k a'_j \right)^n_{j,k=1},$$

that is

$$J_f(q) = \left[e_j - v_j a' \right]^n_{j=1},$$

which is the matrix whose j-th column is the vector

$$e_j - v_j a',$$

where the vector e_j represents the j-th vector of the canonical basis e of $\mathbb{R}^n$ and $v_j a'$ is simply the product of the scalar v_j times the vector a'.

4.3   The Jacobian Determinant

Now we can forwardly calculate the Jacobian determinant in two particular cases, $n = 2$, $n = 3$. We put

$$b_i := -a'_i = (a_i - 1).$$

For $n = 2$, we obtain

$$J_f(q) = \begin{pmatrix} b_1 q_2 + 1 & b_1 q_1 \\ b_2 q_2 & b_2 q_1 + 1 \end{pmatrix} = \begin{pmatrix} b_1 v_1 + 1 & b_1 v_2 \\ b_2 v_1 & b_2 v_2 + 1 \end{pmatrix},$$

and

$$\begin{aligned} \det J_f(q) &= b_1 q_2 + b_2 q_1 + 1 \\ &= b_1 v_1 + b_2 v_2 + 1 \\ &= 1 + (b|v)_2, \end{aligned}$$

where $(b|v)_2$ represents the Euclidean scalar product of b times v in dimension 2. For $n = 3$, we obtain

$$\begin{aligned} J_f(q) &= \begin{pmatrix} b_1 q_2 q_3 + 1 & b_1 q_1 q_3 & b_1 q_1 q_2 \\ b_2 q_2 q_3 & b_2 q_1 q_3 + 1 & b_2 q_1 q_2 \\ b_3 q_2 q_3 & b_3 q_1 q_3 & b_3 q_1 q_2 + 1 \end{pmatrix} \\ &= \begin{pmatrix} b_1 v_1 + 1 & b_1 v_2 & b_1 v_3 \\ b_2 v_1 & b_2 v_2 + 1 & b_2 v_3 \\ b_3 v_1 & b_3 v_2 & b_3 v_3 + 1 \end{pmatrix}, \end{aligned}$$

and

$$\begin{aligned} \det J_f(q) &= b_1 q_2 q_3 + b_2 q_1 q_3 + b_3 q_1 q_2 + 1 \\ &= b_1 v_1 + b_2 v_2 + b_3 v_3 + 1 \\ &= 1 + (b|v)_3. \end{aligned}$$

More difficultly, we can calculate the determinant for the 4-dimensional case. We obtain

$$
J_f(q) = \begin{pmatrix}
b_1 q_2 q_3 q_4 + 1 & b_1 q_1 q_3 q_4 & b_1 q_1 q_2 q_4 & b_1 q_1 q_2 q_3 \\
b_2 q_2 q_3 q_4 & b_2 q_1 q_3 q_4 + 1 & b_2 q_1 q_2 q_4 & b_2 q_1 q_2 q_3 \\
b_3 q_2 q_3 q_4 & b_3 q_1 q_3 q_4 & b_3 q_1 q_2 q_4 + 1 & b_3 q_1 q_2 q_3 \\
b_4 q_2 q_3 q_4 & b_4 q_1 q_3 q_4 & b_4 q_1 q_2 q_4 & b_4 q_1 q_2 q_3 + 1
\end{pmatrix}
$$

$$
= \begin{pmatrix}
b_1 v_1 + 1 & b_1 v_2 & b_1 v_3 & b_1 v_4 \\
b_2 v_1 & b_2 v_2 + 1 & b_2 v_3 & b_2 v_4 \\
b_3 v_1 & b_3 v_2 & b_3 v_3 + 1 & b_3 v_4 \\
b_4 v_1 & b_4 v_2 & b_4 v_3 & b_4 v_4 + 1
\end{pmatrix},
$$

and

$$
\begin{aligned}
\det J_f(q) &= b_1 q_2 q_3 q_4 + b_2 q_1 q_3 q_4 + b_3 q_1 q_2 q_4 + b_4 q_1 q_2 q_3 + 1 \\
&= b_1 v_1 + b_2 v_2 + b_3 v_3 + b_4 v_4 + 1 \\
&= 1 + (b|v)_4.
\end{aligned}
$$

For the geneal case we claim the following result (presented also in [8]).

Theorem 1. *Let $(f, >)$ be the game determined by the strategy space*

$$
E^n = [0, 1]^n
$$

and payoff function defined by

$$
f(q) = \left(q_j - (1 - a_j) \prod_{i=1}^{n} q_i \right)_{j=1}^{n},
$$

for every q belonging to E^n. Then, fixed q belonging to E^n, the Jacobian matrix of the function f at q is the column-defined matrix

$$
J_f(q) = \left[e_j - v_j a' \right]_{j=1}^{n},
$$

where

- *the vector e_j represents the j-th vector of the canonical basis e of $\mathbb{R}^n$;*
- *$v_j a'$ is the product of the scalar v_j times the vector a';*
- *the vector a' shows its i-th component as follows*

$$
a'_i := 1 - a_i,
$$

for every positive integer $i \le n$;

- *the vector*

$$
v := \left(\prod_{i=1, i \ne j}^{n} q_i \right)_{j=1}^{n}
$$

has the component v_j defined as the product

$$v_j = \prod_{i=1, i \neq j}^{n} q_i,$$

for every positive integer $j \leq n$.

Moreover, the determinant of the Jacobian matrix is

$$\det J_f(q) = 1 + (-a'|v),$$

where $(-a'|v)$ is the Euclidean scalr product of $-a'$ times v.

Proof. The first part of the Theorem 1 (regarding the Jacobian matrix) has been already proven in the previous Subsect. 4.2. For what concerns the determinant part, the result follows immediately from the below Lemma 1, by understanding the vector b of the lemma in such a way that

$$J_f(q) = (v_j b + e_j)_{j=1}^n$$

with

$$b = \begin{bmatrix} -a'_1 \\ -a'_2 \\ \vdots \\ -a'_n \end{bmatrix}; \quad v_j = \prod_{i \neq j \, i=1}^{n} q_i; \quad e_j = (\delta_{ij})_{i=1}^n.$$

Now, we state and prove the fundamental lemma for the above Theorem 1.

Lemma 1. *Let b, v be two vectors in $\mathbb{R}^n$ and let*

$$\Lambda_n(b,v) := \det [e_j + v_j b]_{j=1}^n,$$

that is the determinant of the matrix

$$[e_j + v_j b]_{j=1}^n,$$

whose j-th column is the vector

$$e_j + v_j b,$$

where the vector e_j represents the j-th vector of the canonical basis e of $\mathbb{R}^n$ and $v_j b$ is the product of the scalar v_j times the vector b. Then, we obtain

$$\Lambda_n(b,v) = 1 + (b|v)_n,$$

where

$$(b|v)_n = \sum_{i=1}^{n} b_i v_i$$

is the Euclidean scalar product of b times v.

Proof. Let
$$\Lambda_n(b, v) := \Delta[e_j + v_j b]_{j=1}^n,$$

for every positive integer n be the determinants in our analysis, we shall prove our Lemma 1 by recursion on n. The case for $n = 2$ is trivially true. Fixed a positive integer n, let us assume the relation true for the case $n - 1$, we deduce:

$$\Lambda_n(b, v) = \Delta(v_1 b + e_1, v_2 b + e_2, \ldots, v_n b + e_n) \tag{7}$$
$$= \Delta(v_1 b, v_2 b + e_2, \ldots, v_n b + e_n) + \Delta(e_1, v_2 b + e_2, \ldots, v_n b + e_n).$$

We need to calculate the above two determinants. For what concerns the second one, we read

$$\Delta(e_1, v_2 b + e_2, \ldots, v_n b + e_n) = \Lambda_{n-1}(b_*, v_*) = 1 + \sum_{i=2}^n v_i b_i,$$

where:

- the vectors b_*, v_* belong to the $(n - 1)$-Euclidean space and are defined to be the sections of the vectors b, v respectively, by suppressing the first components of both vectors b, v;
- the first equality holds true by the Laplace theorem for the expansion of a determinant;
- the second equality holds true by recursion assumption.

Coming back to the principal formula 7 we finally obtain:

$$\begin{aligned}
\Lambda_n(b, v) &= \Delta(v_1 b + e_1, v_2 b + e_2, \ldots, v_n b + e_n) \\
&= \Delta(v_1 b, v_2 b + e_2, \ldots, v_n b + e_n) + \Lambda_{n-1}(b_*, v_*) \\
&= \Delta(v_1 b, v_2 b, v_3 b + e_3, \ldots, v_n b + e_n) \\
&\quad + \Delta(v_1 b, e_2, v_3 b + e_3, \ldots, v_n b + e_n) + \Lambda_{n-1}(b_*, v_*) \\
&= v_1 v_2 \Delta(b, b, v_3 b + e_3, \ldots, v_n b + e_n) \\
&\quad + \Delta(v_1 b, e_2, v_3 b, v_4 b + e_4, \ldots, v_n b + e_n) \\
&\quad + \Delta(v_1 b, e_2, e_3, v_4 b + e_4, \ldots, v_n b + e_n) + \Lambda_{n-1}(b_*, v_*) \\
&= 0 + v_1 v_3 \Delta(b, e_2, b, v_4 b + e_4, \ldots, v_n b + e_n) \\
&\quad + \Delta(v_1 b, e_2, e_3, v_4 b + e_4, \ldots, v_n b + e_n) + \Lambda_{n-1}(b_*, v_*) \\
&= 0 + 0 + \ldots + 0 + \Delta(v_1 b, e_2, e_3, e_4, \ldots, e_n) + \Lambda_{n-1}(b_*, v_*) \\
&= v_1 b_1 + \Lambda_{n-1}(b_*, v_*) \\
&= v_1 b_1 + 1 + \sum_{i=2}^n v_i b_i \\
&= 1 + (v|b)_n,
\end{aligned}$$

where:

– the determinants

$$\Delta(v_1 b, v_2 b, v_3 b + e_3, \ldots, v_n b + e_n) = v_1 v_2 \Delta(b, b, v_3 b + e_3, \ldots, v_n b + e_n),$$

$$\Delta(v_1 b, e_2, v_3 b, v_4 b + e_4, \ldots, v_n b + e_n) = v_1 v_3 \Delta(b, e_2, b, v_4 b + e_4, \ldots, v_n b + e_n),$$

and so on are all vanishing because each of them shows two proportional columns;
– on the other hand, the last determinant

$$\Delta(v_1 b, e_2, e_3, e_4, \ldots, e_n)$$

equals $v_1 b_1$ because it is the determinant of a low triangular matrix, specifically the below matrix

$$\begin{pmatrix} v_1 b_1 & 0 & 0 & \ldots & 0 \\ v_1 b_2 & 1 & 0 & \ldots & 0 \\ v_1 b_3 & 0 & 1 & \ldots & 0 \\ \vdots & \vdots & \vdots & \ddots & \vdots \\ v_1 b_n & 0 & 0 & \ldots & 1 \end{pmatrix}.$$

So that, the proof is complete.

4.4   Critical Zone

The equation of the critical zone is

$$\det J_f(q) = 0$$

that is

$$1 + \sum_{i=1}^{n} \left(\prod_{j=1, j \neq i}^{n} q_j \right) (a_i - 1) = 0. \tag{8}$$

In this way, we obtain the equation of a hyper-surface representing the *critical zone* of the game. We could prove that a part of the above critical zone is a subset of the maximal Pareto boundary. That information would be useful and necessary to find the Kalai-Smorodinsky solution but we shall follow another way to solve the Kalai decision problem.

4.5   Kalai-Smorodinsky Solution

We write the parametric equations of the straight line connecting, in the n dimensional strategy space, the point a, of components $(a_1, a_2, \ldots, a_n)$, with the payoff supremum point $(1, 1, \ldots, 1)$. We get:

$$\begin{cases} X_1 = a_1 + t(1 - a_1), \\ X_2 = a_2 + t(1 - a_2), \\ \quad \vdots \\ X_n = a_n + t(1 - a_n). \end{cases} \tag{9}$$

where $X = (X_1, \ldots, X_n)$ represents the generic vector payoff of the Kalai-Smorodinsky straight-line. The corresponding locus in the strategy space, inverse image of the straight-line by f, has parametric equations:

$$\begin{cases} f_1(q) = a_1 + t(1 - a_1), \\ f_2(q) = a_2 + t(1 - a_2), \\ \quad \vdots \\ f_n(q) = a_n + t(1 - a_n). \end{cases} \tag{10}$$

By replacing the expressions of $f_1(q)$, $\ldots$, $f_n(q)$, given in 6, we obtain:

$$\begin{cases} q_1 - (1 - a_1) \prod_{i=1}^{n} q_i = a_1 + t(1 - a_1), \\ q_2 - (1 - a_2) \prod_{i=1}^{n} q_i = a_2 + t(1 - a_2), \\ \quad \vdots \\ q_n - (1 - a_n) \prod_{i=1}^{n} q_i = a_n + t(1 - a_n). \end{cases} \tag{11}$$

From the first equation of 11 we obtain

$$\prod_{i=1}^{n} q_i = \frac{q_1 - a_1}{1 - a_1} - t.$$

So that, by eliminating the parameter t, the Eq. 11 are reduced to

$$\begin{cases} q_1 = q_1 \\ q_2 = a_2 + \dfrac{1 - a_2}{1 - a_1}(q_1 - a_1) \\ q_3 = a_3 + \dfrac{1 - a_3}{1 - a_1}(q_1 - a_1) \\ \quad \vdots \\ q_n = a_n + \dfrac{1 - a_n}{1 - a_1}(q_1 - a_1), \end{cases} \tag{12}$$

which are the Cartesian equations of the inverse image of the Kalai-Smorodinsky straight-line.

In order to find the Kalai-Smorodinsky solution S, finally, we shall intersect the above locus with the critical zone of the game: we substitute the expressions 12 in the Eq. 8 (representing the critical zone) and we find an equation of degree $n - 1$ in the argument q_1. If we succeed in finding a q_1, by substitution in 12 we get the entire percentage profile q^* representing the strategy Kalai-Smorodinsky solution of the problem.

By the vector q^*, we immediately find the effectiveness payoff $f(q^*)$ by 6.

4.6   Symmetric Interference

In this subsection, we examine a particular case, relevant in the applications, of symmetric interference among the interacting elements, that is the case in which the vector a shows all the components equal to a particular value $\bar{a}$,

$$a = (\bar{a}, \bar{a}, \ldots, \bar{a}).$$

In this case, we obtain the solution in closed analytical form

$$q^* = (\bar{q}^*, \bar{q}^*, \dots, \bar{q}^*),$$

with

$$\bar{q}^* = \frac{1}{\sqrt[n-1]{n(1 - \bar{a})}}.$$

5 Applications

5.1 An Application to Economics

The problem of finding the optimal quantities of goods to be produced in a certain economy is well established in literature. The circumstance that the demand for certain goods might be influenced by interaction with the demand for other goods often plays a part in this problem. In some cases, a firm needs to decide the production quantities of a product that can partially or completely substitute other products (substitutability). In other cases, the effects of various products may be synergic (complementary).

Let us consider, as a case of other nature, (cannibalization), the case of a company producing two particular commodities (whose quantities are denoted by q_1 and q_2), but which has just developed a new commodity ((whose quantity is denoted by q_3), the demand for which might negatively influence the demands for q_1 and q_2. Let $f_i(q_1, q_2, q_3)$ be the expected market payoff for the i-th product, given the hypothesis in which quantities q_1, q_2 and q_3 of the products are sold. The decision regarding the quantities of the third product to sell depends on the willingness to sacrifice part of the demand for the first two ones. This willingness to cannibalize some products depends on various factors, examples being the future market situation of the three products and a possible company desire to place the third one at a strategic advantage in an emerging market. For a detailed analysis of the factors influencing the willingness to cannibalize see [5, 27, 28]. With the problem defined in these terms, the company can calculate the optimal quantities to produce, applying the method here provided.

5.2 Further Applications

As studied by Carfi et al. [9], this model can be used even in other sectors. For instance, in Public Economics to calibrate differing economic policies that are interfering with each other. There exist also other applications outside economics. In Medicine, the balance of interfering drugs is usually performed by successive approximations, keeping the patient monitored. Thus the decision on the first dose is particularly delicate. Using this model, it is possible to establish the optimal dosages in relation to the desired ratios between improvements in the patient's health with respect to various diseases, taking into account the minimal needed quantity for each medicine.

5.3   Literature Review

In this paper, we shall refer to a wide variety of literature. First of all, we shall consider some papers on the complete study of differentiable games and related mathematical backgrounds, introduced and applied to economic theories since 2008 by Carfi (see [6,7,19]) and by Carfi et al. (see [1,4,20–26]). Specific applications of the previous methodologies, also strictly related to the present model, have been illustrated by Carfi and Musolino in [10–18]. Other important applications of the complete examination methodology were introduced by Carfi and coauthors: ([1,4,9,21–26]).

Finally, we observe that the standard literature on game theory does not present algorithms devoted to the graphical representation and computation of the payoff spaces, but essentially devoted to the determination of Nash equilibria, their stabilities and their approximations.

6   Conclusive Remarks: Features of the Model

We desire to observe some conclusive remarks about our model.

- The model proposed here allows to calculate the optimal quantity percentages of n elements that interfere with each other.
- Our differentiable methods allows to obtain, exactly, the compromise solution vector, in a Kalai-Smorodinsky setting, within a differentiable but not necessarily convex decision problem.
- The extension to the case of payoff functions which are discontinuous or not differentiable on a finite subset of a strategy space, appears possible by fixing few technicalities.
- In our model we go deeply inside the previous two dimensional case, in order to clarify our techniques and some aspects and to prepare the field to the n-dimensional case, that constitutes the core of this paper.
- The compromise solution, that is the solution of our decision problem, is determined, fixed the dimension of the problem, by an original algorithm which can be applied with the chosen computational precision.
- The algorithm was written with the applicative software Matlab, obviously based on the analytical methods proposed by Carfi in [7].

7   Conclusions

The study proposed here:

- builds up a specific mathematical game model for n interacting elements;
- provides algorithms associated with the model;
- shows the optimal Kalai-Smorodinsky compromise solution n-tuples of active principle quantity percentages in the game;
- examines a non-linear scenario with n possible interacting elements;
- the core section of the paper, completely studies the game in n-dimensions;

- we find the critical zone of the game in its Cartesian form;
- we prove a theorem and a lemma about the Jacobian determinant of the n-game;
- we write down the intersection of the critical zone and the Kalai-Smorodinsky straight-line and we solve numerically the problem in every concave case;
- in a particular highly symmetrical case, we solve analytically the Kalai-Smorodinsky compromise problem.

We solve the problem by using;

- the Carfì's payoff analysis method for differentiable payoff functions;
- algorithms for the finding of Kalai-Smorodinsky solutions;
- softwares for the determination of graphs in the two dimensional case.

Appendix 1: The Algorithm

Here the Matlab command list follows.

Kalai-Smorodinsky Solution in 2 Dimension

```
syms ('p', 'q', 'a', 'b')
f1 =  p*(1−q)+ a*p*q;
f2 =  q*(1−p)+b*p*q;
f =  [f1; f2];
v =  [p q];
J =  jacobian (f, v)
D =  det (J)
R =  b+((1−b)./(1−a)).*(p−a)−q
K =  solve (D, R, p, q)
p =  K.p
q =  K.q
```

which provides:

```
J =
[ a*q − q + 1,     a*p − p]
[       b*q − q,  b*p − p + 1]

D =   a*q − q − p + b*p + 1

R =   b − q − ((a − p)*(b − 1))/(a − 1)

K =
      p: [1 x1  sym]
      q: [1 x1  sym]
```

$$p = -(a - b + 1)/(2*b - 2)$$

$$q = -(b - a + 1)/(2*a - 2)$$

Numerical Sample in the Case $n = 10$

```
a1  =  0;
a2  =  0.01;
a3  =  0.03;
a4  =  0.032;
a5  =  0.034;
a6  =  0.04;
a7  =  0.041;
a8  =  0.043;
a9  =  0.048;
a10  =  0.3;
```

```
syms ('q1', 'q2', 'q3', 'q4', 'q5', 'q6', 'q7', 'q8', 'q9', 'q10')
f1  =  q1  -(1-a1)* q1* q2* q3* q4* q5* q6* q7* q8* q9* q10;
f2  =  q2  -(1-a2)* q1* q2* q3* q4* q5* q6* q7* q8* q9* q10;
f3  =  q3  -(1-a3)* q1* q2* q3* q4* q5* q6* q7* q8* q9* q10;
f4  =  q4  -(1-a4)* q1* q2* q3* q4* q5* q6* q7* q8* q9* q10;
f5  =  q5  -(1-a5)* q1* q2* q3* q4* q5* q6* q7* q8* q9* q10;
f6  =  q6  -(1-a6)* q1* q2* q3* q4* q5* q6* q7* q8* q9* q10;
f7  =  q7  -(1-a7)* q1* q2* q3* q4* q5* q6* q7* q8* q9* q10;
f8  =  q8  -(1-a8)* q1* q2* q3* q4* q5* q6* q7* q8* q9* q10;
f9  =  q9  -(1-a9)* q1* q2* q3* q4* q5* q6* q7* q8* q9* q10;
f10  =  q10  -(1-a10)* q1* q2* q3* q4* q5* q6* q7* q8* q9* q10;

f  =  [f1;  f2;  f3;  f4;  f5;  f6;  f7;  f8;  f9;  f10];
v  =  [q1  q2  q3  q4  q5  q6  q7  q8  q9  q10];
J  =  jacobian (f, v);
D  =  det (J);
k  =  D;

A  =  subs (k,  q2,  (a2+(1-a2)/(1-a1)*(q1-a1)));
B  =  subs (A,  q3,  (a3+(1-a3)/(1-a1)*(q1-a1)));
C  =  subs (B,  q4,  (a4+(1-a4)/(1-a1)*(q1-a1)));
E  =  subs (C,  q5,  (a5+(1-a5)/(1-a1)*(q1-a1)));
F  =  subs (E,  q6,  (a6+(1-a6)/(1-a1)*(q1-a1)));
G  =  subs (F,  q7,  (a7+(1-a7)/(1-a1)*(q1-a1)));
H  =  subs (G,  q8,  (a8+(1-a8)/(1-a1)*(q1-a1)));
I  =  subs (H,  q9,  (a9+(1-a9)/(1-a1)*(q1-a1)));
L  =  subs (I,  q10,  (a10+(1-a10)/(1-a1)*(q1-a1)));
```

$K = solve(L, q1)$

$q1 = K(1)$
$q2 = a2+(1-a2)/(1-a1)*(K(1)-a1)$
$q3 = a3+(1-a3)/(1-a1)*(K(1)-a1)$
$q4 = a4+(1-a4)/(1-a1)*(K(1)-a1)$
$q5 = a5+(1-a5)/(1-a1)*(K(1)-a1)$
$q6 = a6+(1-a6)/(1-a1)*(K(1)-a1)$
$q7 = a7+(1-a7)/(1-a1)*(K(1)-a1)$
$q8 = a8+(1-a8)/(1-a1)*(K(1)-a1)$
$q9 = a9+(1-a9)/(1-a1)*(K(1)-a1)$
$q10 = a10+(1-a10)/(1-a1)*(K(1)-a1)$

$f1 = q1 -(1-a1)*q1*q2*q3*q4*q5*q6*q7*q8*q9*q10$
$f2 = q2 -(1-a2)*q1*q2*q3*q4*q5*q6*q7*q8*q9*q10$
$f3 = q3 -(1-a3)*q1*q2*q3*q4*q5*q6*q7*q8*q9*q10$
$f4 = q4 -(1-a4)*q1*q2*q3*q4*q5*q6*q7*q8*q9*q10$
$f5 = q5 -(1-a5)*q1*q2*q3*q4*q5*q6*q7*q8*q9*q10$
$f6 = q6 -(1-a6)*q1*q2*q3*q4*q5*q6*q7*q8*q9*q10$
$f7 = q7 -(1-a7)*q1*q2*q3*q4*q5*q6*q7*q8*q9*q10$
$f8 = q8 -(1-a8)*q1*q2*q3*q4*q5*q6*q7*q8*q9*q10$
$f9 = q9 -(1-a9)*q1*q2*q3*q4*q5*q6*q7*q8*q9*q10$
$f10 = q10 -(1-a10)*q1*q2*q3*q4*q5*q6*q7*q8*q9*q10$

which provides:

$K =$
$$
\begin{array}{l}
0.7659 \\
0.7119*i - 0.4900 \\
- 0.2802*i - 0.8617 \\
- 0.7119*i - 0.4900 \\
0.8120*i + 0.0753 \\
0.0753 - 0.8120*i \\
- 0.5308*i + 0.5705 \\
0.5308*i + 0.5705 \\
0.2802*i - 0.8617
\end{array}
$$

$q1 = 0.7659$
$q2 = 0.7682$
$q3 = 0.7729$
$q4 = 0.7734$
$q5 = 0.7739$
$q6 = 0.7753$
$q7 = 0.7755$
$q8 = 0.7760$
$q9 = 0.7771$
$q10 = 0.8361$

$f1 = 0.6834$
$f2 = 0.6865$
$f3 = 0.6929$
$f4 = 0.6935$
$f5 = 0.6941$
$f6 = 0.6960$
$f7 = 0.6964$
$f8 = 0.6970$
$f9 = 0.6986$
$f10 = 0.7784$

Appendix 2: Differentiable Games and Pareto Boundary

In this paper we use a general method introduced by one of the authors in [7] and referred in literature as Complete Study of a Differentiable Game. Indeed, in the great part of the current Game Theory literature, the study of a game in normal form consists essentially in the determination of the Nash equilibria (in mixed strategies) and in the analysis of their stability properties (see for instance [2,3,29,30]). This approach cannot provide a complete and global view of the game and of all its possible feasible solutions. Indeed, to figure out how the game determines the interaction among the players, a deeper knowledge of at least the payoff space of the game appears not only useful but mandatory. For instance, it appears of the greatest interest to know the positions of the payoff profiles corresponding to the Nash equilibria in the whole of the payoff space of the game and the knowledge of these relative locations requires the knowledge of the entire payoff space. The need of better knowing the general shape of the payoff profile space becomes inevitable, when the problem to solve in the game reveals a bargaining one, at least at the level of the maximal boundary of the payoff space: in fact, the exact determination of bargaining solutions (compromise solutions) needs the analytical determination of the maximal Pareto boundary. In the paper [7], Carfi presented a general method to find an explicit expression of the topological boundary of the payoff space of the game. In that paper Carfi followed the way shown in [6], in order to construct the theoretical bases for Decisions in Economics and Finance by means of algebraic, topological and differentiable structures.

Preliminaries and Notations

Here we reconsider, for convenience of the reader, the study conducted in the paper [23]. In order to help the reader and increase the level of readability of the paper, we recall some notations and definitions about n-player games in normal-form, presented yet in [1,7]. Although the below definition seems, at a first sight, different from the standard one (presented, for example, in [29]), we desire to note that it is substantially the same; on the other hand, the definition in this new form underlines that a normal-form game is nothing but a vector-valued function and that any possible exam or solution of a normal-form games attains, indeed, to this functional nature.

Definition 1 (definition of a game in normal form). *Let $E = (E_i)_{i=1}^n$ be an ordered family of non-empty sets. We call n-**person game in normal-form, upon the support** E, each function $f : {}^\times E \to \mathbb{R}^n$, where ${}^\times E$ denotes the Cartesian product $\times_{i=1}^n E_i$ of the family E. The set E_i is called the **strategy set of player** i, for every index i of the family E, and the product ${}^\times E$ is called the **strategy profile space**, or the n-strategy space, of the game.*

Remark 2. First of all we recall a standard form definition of normal-form game.

Definition 2. *A strategic game consists of a system (N, E, f), where:*

1. *a finite set N (the set of players) of cardinality n is canonically identified with the set of the first n positive integers;*
2. *E is an ordered family of nonempty sets, $E = (E_i)_{i \in N}$, where, for each player i in N, the nonempty set E_i is the set of actions available to player i;*
3. *f is an ordered family of real functions $f = (f_i)_{i \in N}$, where, for each player i in N, the function $f_i : {}^\times E \to \mathbb{R}$ is the utility function of player i (inducing a preference relation on the Cartesian product ${}^\times E := \times_{j \in N} E_j$ (the preference relation of player i on the whole strategy space).*

Well, it is quite clear that the above system (N, E, f) is nothing but a redundant form of the family f itself, which we prefer to consider in its vector-valued functional nature

$$f : \times_{j \in N} E_j \to \mathbb{R}^n : x \mapsto (f_i(x))_{i \in N}.$$

Terminology. Together with the previous definition of game in normal form, we need to introduce some terminologies:

- the set $\{i\}_{i=1}^n$ of the first n positive integers is said *the set of players* of the game;
- each element of the Cartesian product ${}^\times E$ is said a *strategy profile*, or n-strategy, of the game;
- the image of the function f, i.e., the set $f({}^\times E)$ of all real n-vectors of type $f(x)$, with x in the strategy profile space ${}^\times E$, is called the *n-payoff space*, or simply the *payoff space*, of the game f.

Moreover, we recall the definition of Pareto boundary whose main properties have been presented in [6]. By the way, the maximal boundary of a subset T of the Euclidean space $\mathbb{R}^n$ is the set of those $s \in T$ which are not strictly less than any other element of T.

Definition 3 (Pareto boundary). *The Pareto maximal boundary of a game f is the subset of the n-strategy space of those n-strategies x such that the corresponding payoff $f(x)$ is maximal in the n-payoff space, with respect to the usual* *order of the euclidean n-space* $\mathbb{R}^n$. *If* S *denotes the strategy space* $^\times E$, *we shall denote the maximal boundary of the n-payoff space by* $\overline{\partial}f(S)$ *and the maximal boundary of the game by* $\overline{\partial}_f(S)$ *or by* $\overline{\partial}(f)$. *In other terms, the maximal boundary* $\overline{\partial}_f(S)$ *of the game is the reciprocal (inverse) image (by the function* f) *of the maximal boundary of the payoff space* $f(S)$. *We shall use analogous terminologies and notations for the* minimal *Pareto boundary.*

Remark 3 (on the definition of Pareto boundary). Also in the case of this definition, essentially the definition of maximal (Pareto) boundary is the standard one, unless perhaps the name Pareto: it is nothing more that the set of maximal elements in the standard pre-order set sense, that is the set of all elements that are not strictly less than other elements of the set itself. The only circumstance to point out is that the natural pre-order of the strategy set $^\times E$ is that induced by the standard point-wise order of the image $f(S)$ by means of the function f, that is the reciprocal image (Bourbaki's term for inverse image) of the point-wise order on $f(S)$ via f.

The Method for C^1 Games. In this paper, we deal with normal-form game f defined on the product of n compact and non-degenerate intervals of the real line, and such that f is the restriction to the n-strategy space of a C^1 function defined on an open set of $\mathbb{R}^n$ containing the n-strategy space S (which, in this case, is a compact infinite part of the n-space $\mathbb{R}^n$). Details can be found in [7,20,21] but in the following we recall some basic notions.

Topological Boundary. The key theorem of our method is the following one, we invite the reader to **pay much attention to the topologies used below.**

Theorem 2. *Let* f *be a* C^1 *function defined upon an open set* O *of the euclidean space* $\mathbb{R}^n$ *and with values in* $\mathbb{R}^n$. *Then, for every part* S *of the open set* O, *the topological boundary of the image of* S *by the function* f, **in the topological space** $f(O)$ *(i.e. with respect to the relativization of the Euclidean topology to* $f(O)$) *is contained in the union*

$$f(\partial_O S) \cup f(C),$$

that is

$$\partial_{f(O)} f(S) \subseteq f(\partial_O S) \cup f(C),$$

where:

1. C *is the critical set of the function* f *in* S *(that is the set of all points* x *of* S *such that the Jacobian matrix* $J_f(x)$ *is not invertible);*
2. $\partial_O S$ *is the topological boundary of* S *in* O *(with respect to the relative topology of* O).

We strongly invite the reader to see the definitions and remarks about and around Theorem 2 in [7,20,21,23].

References

1. Agreste, S., Carfi, D., Ricciardello, A.: An algorithm for payoff space in C^1 parametric games. Appl. Sci. **14**, 1–14 (2012). http://www.mathem.pub.ro/apps/v14/A14-ag.pdf
2. Aubin, J.: Optima and Equilibria. Springer, Heidelberg (1993)
3. Aubin, J.: Mathematical Methods of Game and Economic Theory. Dover Publications, New York (2007)
4. Baglieri, D., Carfi, D., Dagnino, G.: Asymmetric R&D alliances and coopetitive games. In: Greco, S., Bouchon-Meunier, B., Coletti, G., Fedrizzi, M., Matarazzo, B., Yager, R. (eds.) IPMU 2012. CCIS, vol. 300, pp. 607–621. Springer, Heidelberg (2012). https://doi.org/10.1007/978-3-642-31724-8_64
5. Battagion, M.R., Grieco, D.: R&D competition with radical and incremental innovation. Rivista Italiana degli Economisti 2 (2009)
6. Carfi, D.: Optimal boundaries for decisions. AAPP — Phys. Math. Nat. Sci. **86**(1), 1–11 (2008). https://dx.doi.org/10.1478/C1A0801002
7. Carfi, D.: Payoff space in C^1 Games. Appl. Sci. (APPS) **11**, 35–47 (2009). http://www.mathem.pub.ro/apps/v11/A11-ca.pdf
8. Carfi, D.: A game Pareto complete analysis in n-dimensions: a general applicative study case. J. Math. Econ. Finance (2017), forthcoming in Summer Issue
9. Carfi, D., Gambarelli, G., Uristani, A.: Balancing pairs of interfering elements. Zeszyty Naukowe Uniwersytetu Szczecińskiego - Finanse, Rynki Finansowe, Ubezpieczenia **760**(59), 435–442 (2013)
10. Carfi, D., Musolino, F.: A coopetitive approach to financial markets stabilization and risk management. In: Greco, S., Bouchon-Meunier, B., Coletti, G., Fedrizzi, M., Matarazzo, B., Yager, R.R. (eds.) IPMU 2012. CCIS, vol. 300, pp. 578–592. Springer, Heidelberg (2012). https://doi.org/10.1007/978-3-642-31724-8_62
11. Carfi, D., Musolino, F.: Game theory and speculation on government bonds. Econ. Model. **29**(6), 2417–2426 (2012). https://dx.doi.org/10.1016/j.econmod.2012.06.037
12. Carfi, D., Musolino, F.: Credit crunch in the euro area: a coopetitive multi-agent solution. In: Ventre, A.G.S., Maturo, A., Hoškovà-Mayerovà, Š., Kacprzyk, J. (eds.) Multicriteria and Multiagent Decision Making with Applications to Economics and Social Sciences, STUDFUZZ, vol. 305, pp. 27–48. Springer, Heidelberg (2013). https://dx.doi.org/10.1007/978-3-642-35635-3_3
13. Carfi, D., Musolino, F.: Game theory application of Monti's proposal for European government bonds stabilization. Appl. Sci. **15**, 43–70 (2013). http://www.mathem.pub.ro/apps/v15/A15-ca.pdf
14. Carfi, D., Musolino, F.: Model of possible cooperation in financial markets in presence of tax on speculative transactions. AAPP — Phys. Math. Nat. Sci. **91**(1), 1–26 (2013). https://dx.doi.org/10.1478/AAPP.911A3
15. Carfi, D., Musolino, F.: Dynamical stabilization of currency market with fractal-like trajectories. Sci. Bull. Politehnica Univ. Bucharest **76**(4), 115–126 (2014). http://www.scientificbulletin.upb.ro/rev_docs_arhiva/rezc3a_239636.pdf
16. Carfi, D., Musolino, F.: Speculative and hedging interaction model in oil and U.S. dollar markets with financial transaction taxes. Econ. Model. **37**, 306–319 (2014). https://dx.doi.org/10.1016/j.econmod.2013.11.003
17. Carfi, D., Musolino, F.: A coopetitive-dynamical game model for currency markets stabilization. AAPP — Phys. Math. Nat. Sci. **93**(1), 1–29 (2015). https://dx.doi.org/10.1478/AAPP.931C1

18. Carfì, D., Musolino, F.: Tax evasion: a game countermeasure. AAPP — Phys. Math. Nat. Sci. **93**(1), 1–17 (2015). https://dx.doi.org/10.1478/AAPP.931C2
19. Carfì, D., Ricciardello, A.: Non-reactive strategies in decision-form games. AAPP — Phys. Math. Nat. Sci. **87**(2), 1–12 (2009). https://dx.doi.org/10.1478/C1A0902002
20. Carfì, D., Ricciardello, A.: An algorithm for payoff space in C^1-Games. AAPP — Phys. Math. Nat. Sci. **88**(1), 1–19 (2010). https://dx.doi.org/10.1478/C1A1001003
21. Carfì, D., Ricciardello, A.: Algorithms for payoff trajectories in C^1 parametric games. In: Greco, S., Bouchon-Meunier, B., Coletti, G., Fedrizzi, M., Matarazzo, B., Yager, R.R. (eds.) IPMU 2012. CCIS, vol. 300, pp. 642–654. Springer, Heidelberg (2012). https://doi.org/10.1007/978-3-642-31724-8_67
22. Carfì, D., Ricciardello, A.: Topics in Game Theory. Applied Sciences - Monographs 9. Balkan Society of Geometers (2012). http://www.mathem.pub.ro/apps/mono/A-09-Car.pdf
23. Carfì, D., Ricciardello, A.: An algorithm for dynamical games with fractal-like trajectories. In: Carfì, D., Lapidus, M., Pearse, E., Van Frankenhuijsen, M. (eds.) Fractal Geometry and Dynamical Systems in Pure and Applied Mathematics II: Fractals in Applied Mathematics. (PISRS 2011 International Conference on Analysis, Fractal Geometry, Dynamical Systems and Economics, Messina, Italy, November 8–12, 2011 - AMS Special Session on Fractal Geometry in Pure and Applied Mathematics: in Memory of Benoît Mandelbrot, Boston, Massachusetts, 4–7 January, 2012 - AMS Special Session on Geometry and Analysis on Fractal Spaces, Honolulu, Hawaii, 3–4 March, 2012), Contemporary Mathematics, vol. 601, pp. 95–112. American Mathematical Society (2013). https://dx.doi.org/10.1090/conm/601/11961
24. Carfì, D., Schilirò, D.: A Framework of coopetitive games: applications to the Greek crisis. AAPP — Phys. Math. Nat. Sci. **90**(1), 1–32 (2012). https://dx.doi.org/10.1478/AAPP.901A1
25. Carfì, D., Schilirò, D.: A coopetitive model for the green economy. Econ. Model. **29**(4), 1215–1219 (2012). https://dx.doi.org/10.1016/j.econmod.2012.04.005
26. Carfì, D., Schilirò, D.: Global green economy and environmental sustainability: a coopetitive model. In: Greco, S., Bouchon-Meunier, B., Coletti, G., Fedrizzi, M., Matarazzo, B., Yager, R.R. (eds.) IPMU 2012. CCIS, vol. 300, pp. 593–606. Springer, Heidelberg (2012). https://doi.org/10.1007/978-3-642-31724-8_63
27. Chandy, R.K., Tellis, G.J.: Organizing for radical product innovation: the overlooked role of willingness to cannibalize. J. Mark. Res. **35**, 474–487 (1998)
28. Nijssen, E.J., Hillebrand, B., Vermeulen, P.A.M., Kemp, R.: Understanding the Role of Willingness to Cannibalize in New Service Development. Scales Research Reports H200308 (2004)
29. Osborne, M.J., Rubinstein, A.: A Course in Game Theory. Academic Press, Vancouver (2001)
30. Owen, G.: Game Theory. Academic Press, London (2001)

Reason vs. Rationality: From Rankings to Tournaments in Individual Choice

Janusz Kacprzyk[1], Hannu Nurmi[2(✉)], and Sławomir Zadrożny[1]

[1] Systems Research Institute, Polish Academy of Sciences, Warsaw, Poland
[2] Department of Philosophy, Contemporary History and Political Science, University of Turku, Turku, Finland
hnurmi@utu.fi

Abstract. The standard assumption in decision theory, microeconomics and social choice is that individuals (consumers, voters) are endowed with preferences that can be expressed as complete and transitive binary relations over alternatives (bundles of goods, policies, candidates). While this may often be the case, we show by way of toy examples that incomplete and intransitive preference relations are not only conceivable, but make intuitive sense. We then suggest that fuzzy preference relations and solution concepts based on them are plausible in accommodating those features that give rise to intransitive and incomplete preferences. Tracing the history of those solutions leads to the works of Zermelo in 1920's.

1 Introduction

A basic concept in decision theory is that of rationality. While no decision theorist would maintain that all human behaviour is rational, most of them would probably argue that rational behaviour provides a useful benchmark for evaluating and explaining any kind of behaviour. In particular, if observed behaviour is found to agree with the dictates of rationality, no further explanation is typically needed for it. It is behaviours that exhibit deviations from rationality that require explanation. But what is then rational behaviour? The most precise definition – due to Savage (1954) – is based on a simple choice situation involving two alternatives, say, A and B. Suppose that the individual making the choice has a strict preference over these two so that he/she (hereinafter he) strictly prefers A to B. Choice behaviour is then called rational if it always, that is, with probability 1, results in A being chosen (see also Harsanyi 1977). Of course, we may encounter situations where the individual is physically prevented from choosing A or by making him believe that A is not really available or that by taking some new aspects of the situation into account, he does not prefer A to B or something similar. These kinds of considerations are, however, irrelevant since by suggesting that A is not available, the situation is no longer one involving a choice. Similarly, if the individual is led to believe that he is actually preferring

---

The authors are grateful to the referees for perceptive and constructive comments on an earlier version.

N.T. Nguyen et al. (Eds.): TCCI XXVII, LNCS 10480, pp. 28–39, 2017.
https://doi.org/10.1007/978-3-319-70647-4_2

B to A, the "original" preference no longer holds. So, we may argue that the definition holds at least as far as *preference-based rationality* is concerned. In this setting it is quite straight-forward and trivial to argue that rational behaviour aims at utility maximization since by assigning a larger utility value to A than to B, we guarantee that preferences coincide with utility maximization.

Things get more complicated when the alternative set is expanded. The standard way to proceed is to impose conditions on preference relations that guarantee that acting in accordance with preferences amounts to maximizing utilities. In other words, one looks for properties that the preferences have to possess in order for the choices made according to those preferences to be equivalent to utility maximization by the chooser. In fact, the theory of choice under certainty, risk and uncertainty focuses precisely on those conditions. The standard representation theorem (see e.g. Harsanyi 1977, 31) states that if the individual is endowed with a continuous, complete and transitive weak preference relation over the alternatives, then his choice behaviour – if it conforms with his preferences – can be represented as utility maximizing.[1] In the following sections we shall consider each one of these properties of preference relations in turn and discuss their plausibility. Our aim is to show that under relatively general circumstances each one of them can be questioned. We shall thereafter endeavour to show that fuzzy binary preference relations could provide a useful starting point for modelling reason-based behaviour and a more plausible benchmark than the traditional preference-based rationality.

2   Transitivity Assumption

It is common to assume that preferences are revealed by choices. This is, in fact, stated in the definition of preference. In the world of empirical observations it may, however, happen that a person may, for one reason or another, occasionally choose B even though his preference is for A over B. It would, then, be more plausible to translate the preference of A over B into a probability statement according to which the probability of A being chosen by the person is larger than the probability that B is chosen. Starting from this somewhat milder probability definition of preference, we shall now consider the transitivity property. May (1954) suggests that the appropriate definition of preference-based choice is one that – in addition to choice probability – includes the alternative set considered as well as the description of the experimental setting. In this framework the preference for A over B is expressed as the following probability statement:

$$p(A|A, B, E) > p(B|A, B, E)$$

Here E denotes the experimental setup.

Suppose now that A is preferred to B and B is preferred to C. I.e.

$$p(A|A, B, E) > p(B|A, B, E) \tag{1}$$

---

[1] The weak preference of A over B means that A is regarded as at least as desirable as B. Thus the weak preference relation is not asymmetric, while the strong one is.

$$p(B|B, C, E) > p(C|B, C, E) \qquad (2)$$

Now, transitivity would require that Eqs. 1 and 2 imply that

$$p(A|A, C, E) > p(C|A, C, E) \qquad (3)$$

It is, however, difficult to associate this implication with rationality, since the alternative sets considered are different in each Equation: in Eq. 1 it is $\{A, B\}$, in Eq. 2 it is $\{B, C\}$ and in Eq. 3 it is $\{A, C\}$. What May (1954, 2) argues is "that transitivity does not follow from this empirical [probabilistic] interpretation of preference, but must be established, if at all, by empirical observation." This point on which we completely agree leaves, however, open the possibility that transitivity would be normatively compelling (even if empirically contestable). Our position is stronger here: while we agree that there are circumstances where transitivity seems normatively plausible[2], there are others where it is not. Hence, defining rationality so that transitivity of preferences is a necessary part of it, is not acceptable in our view.

The reason is rather straight-forward. The grounds for preferring A over B might well be different from those used in ranking B ahead of C. Hence, it is purely contingent whether these or other grounds are used in preferring C to A or vice versa. Alternatively, the decision maker may use several criteria of "performance" of alternatives. Each of these may result in a complete and transitive relation over alternatives, but when forming the overall preference relation on the basis of these rankings, the decision maker may well end up with an intransitive relation. Consider a fictitious example.

Three universities A, B and C are being compared along three criteria: (i) research output (scholarly publications), (ii) teaching output (degrees), (iii) external impact (expert assignments, media visibility, R&D projects, etc.)

Publications	Teaching	External impact
A	B	C
B	C	A
C	A	B

Assuming that each criterion is of roughly equal importance, it is natural to form the overall preference relation between the universities on the basis of the majority rule: which one of any two universities is ranked higher than the other by a majority of criteria is preferred to the latter. In the present example this leads to a cycle: $A \succ B \succ C \succ A \succ \ldots$. Hence, intransitive individual preference relations can be made intelligible by multiple criterion setting and majority principle (cf. Fishburn 1970; Bar-Hillel and Margalit 1988).

---

[2] E.g. in preferences over monetary payoffs.

3   Completeness Assumption

The completeness of weak preference relation entails that for any pair (A, B) of alternatives either A is preferred to B or vice versa or both. Stated in another way, completeness means that it cannot be the case that A is not preferred to B and B is not preferred to A. In the following we show that there is nothing unnatural or irrational in situations where there are grounds for saying that neither A is preferred to B nor B is preferred to A. Perhaps the simplest way to show this is via a phenomenon known as Ostrogorski's paradox. It refers to the ambiguity in determining the popular preference among two alternatives (Daudt and Rae 1978). In the following we recast this paradox in an individual decision-making setting. The nominating individual is to make a choice between two alternatives A and B, e.g. applicants to the chair of economics in a university. Three types of merits are deemed of primary importance for this office, viz. research merits, teaching skills and ability to attract external funding to the university. The nominating individual received advice from three other individuals: one representing the peers (i.e. other economics professors), one representing the students of economics and one representing the university administration. The following table indicates the preferred applicant of each representative on each area of merit. Thus, e.g. applicant A has a preferable research record according to the peers than applicant B. Similarly, the representative of the administration deems B preferable in each merit area.[3]

Merit area	Research	Teaching	Funding potential	Row choice
Advisor 1	A	B	A	A
Advisor 2	A	A	B	A
Advisor 3	B	B	B	B
Column Choice	A	B	B	?

Suppose now that the nominating individual forms his preference in a neutral and anonymous manner, i.e. each merit area and each advisor is considered equally important. It would then appear natural that whichever applicant is deemed more suitable by more advisors than its competitor, is preferable in the respective merit area. Similarly, whichever candidate is more suitable than his competitor in more merit areas, is regarded as preferable by each advisor. Under these assumptions the nominating individual faces a quandary: if the aggregation of valuations is first done over columns – i.e. each advisor's overall preference is determined first – and then over rows – i.e. picking the applicant regarded more appropriate by the majority of advisors – the outcome is that B cannot be

---

[3] All preferences underlying the table are assumed to be strict. The composition of the advisory body may raise some eyebrows. So, instead of these particular categories of advisors, one may simply think of a body that consists of three peers.

preferred to A. If the aggregations is performed in the opposite order – first over rows and then over columns – the outcome is that A cannot be preferred to B. Hence, the preference relation over $\{A, B\}$ is not complete.

It should be observed that there is nothing arbitrary or irrational in the above example. The use of expert information (advisors) or other evaluation criteria in assessing applicants would seem quite natural way to proceed. Also, the task to be performed by the successful applicant often has several aspects (merit areas) to it. Similarly, the use of majority principle in determining the "winners" of aggregation is quite reasonable, certainly not counterintuitive.

4 Continuity Assumption

Continuity axiom states that both the inferior and superior sets for any given alternative are closed (Harsanyi 1977, 31). To elaborate this a little, consider a set $\mathcal{X}$ of alternatives and an element x in it. Let now $x_1, x_2, \ldots$, a sequence of alternatives converging to x_0, have the property that for each x_i in the sequence, $x \succ_j x_i$. In other words, individual j prefers x to each element of the sequence. Then, continuity requires that $x \succ_j x_0$ as well. Similarly, continuity requires that the sequence has the property that if $x_i \succ_j x$ for each x_i in the sequence, then $x_0 \succ_j x$ as well. The above pertains to infinite sequences. In finite ones, continuity requires that small changes in the alternatives are accompanied with small changes in their desirability.

Let us now see how continuity assumption translates into multiple-criterion settings. We shall take advantage of Baigent (1987) fundamental result in social choice theory. This result has subsequently been augmented, modified and generalized by Eckert and Lane (2002), Baigent and Eckert (2004), as well as by Baigent and Klamler (2004). We shall, however, largely make use of the early version (Baigent 1987). It states the following.

Theorem 1. *Anonymity and respect for unanimity of a social choice function cannot be reconciled with proximity preservation.*

Proximity preservation is a property defined for social choice functions. It amounts to the requirement that choices made in profiles more close to each other ought to be closer to each other than those made in profiles less close to each other. Profiles – it will be recalled – are n-tuples of preference rankings over the set of alternatives (n being the number of individuals). What this requirement intuitively means is that if we make a small modification in the preference rankings, the change in the outcome of the social choice function should be smaller than if we make a larger modification. Anonymity, in turn, requires that relabelling of the individuals does not change the choice outcomes. In multi-criterion setting anonymity means that permuting the criteria does not change the outcome of evaluation. Respect for unanimity is satisfied whenever the choice function agrees with a preference ranking held by all individuals, i.e. if $x \succ_j y$ for all individuals j, then this will also be the social ranking between x and y. In multi-criterion environment this amounts to the requirement that if all criteria suggest the same ranking of alternatives, then this ranking should also be the outcome.

To illustrate the incompatibility exhibited by Baigent's theorem, let us turn again to the fictitious example of nominating the chair of economics. Suppose that there are two applicants A and B. Moreover, only two criteria are being used by the nominating authority: research merits (R) and teaching record (T).[4] To simplify things, assume that only strict preferences are possible, i.e. each criterion produces a strict ranking of the applicants. Four different configurations of rankings (i.e. profiles) $(S_1, \ldots, S_4)$ are now possible:

S_1		S_2		S_3		S_4	
R	T	R	T	R	T	R	T
A	A	B	B	B	A	A	B
B	B	A	A	A	B	B	A

Let us denote the rankings in various configurations S_{mi} where m is the number of the configuration and i the criterion. We consider two types of metrics: one that is defined on pairs of rankings and one defined on configurations. The former is denoted by d_r an the latter by d_S. The number of criteria is N. The metrics are related as follows:

$$d_S(S_m, S_j) = \sum_{i \in N} d_r(S_{mi}, S_{ji}).$$

In other words, the distance between two configurations is the sum of distances between the pairs of rankings of the first, second, *etc.* criterion.

Take now two configurations, S_1 and S_3, from the above list and express their distance using metric d_S as follows:

$$d_S(S_1, S_3) = d_r(S_{11}, S_{31}) + d_r(S_{12}, S_{32}).$$

Since, $S_{12} = S_{32} = A \succ B$, and hence the latter summand equals zero, this reduces to:

$$d_S(S_1, S_3) = d_r(S_{11}, S_{31}) = d_r((A \succ B), (B \succ A)).$$

Taking now the distance between S_3 and S_4, we get:

$$d_S(S_3, S_4) = d_r(S_{31}, S_{41}) + d_r(S_{32}, S_{42}).$$

Both summands are equal since by definition:

$$d_r((B \succ A), (A \succ B)) =$$
$$d_r((A \succ B), (B \succ A)).$$

Thus,

---

[4] The argument is a slight modification of Baigent's (1987, 163) illustration.

$$d_S(S_3, S_4) = 2 \times d_r((A \succ B), (B \succ A)).$$

In terms of d_S, then, S_3 is closer to S_1 than to S_4. This makes sense intuitively.

We now turn to procedures used in aggregating the information on criterion-wise rankings into an overall evaluation or choice. Let us denote the aggregation procedure by g. We make two intuitively plausible restrictions on choice procedures, *viz.* that they are anonymous and respect unanimity. In our example, anonymity requires that whatever is the choice in S_3 is also the choice in S_4 since these two profiles can be reduced to each other by relabelling the criteria. Unanimity, in turn, requires that $g(S_1) = A$, while $g(S_2) = B$. Therefore, either $g(S_3) \neq g(S_1)$ or $g(S_3) \neq g(S_2)$. Assume the former. It then follows that $d_r(g(S_3), g(S_1)) > 0$. Recalling the implication of anonymity, we now have:

$$d_r(g(S_3), g(S_1)) > 0 = d_r(g(S_3), g(S_4)).$$

In other words, even though S_3 is closer to S_1 than to S_4, the choice made in S_3 is closer to - indeed identical with - that made in S_4. This argument rests on the assumption that $g(S_3) \neq g(S_1)$. Similar argument can, however, easily be made for the alternative assumption, *viz.* that $g(S_3) \neq g(S_2)$.

The example shows that small mistakes or errors in criterion measurements are not necessarily accompanied with small changes in evaluation outcomes. Indeed, if the true criterion rankings are those of S_3, then a mistaken report on criterion 1's leads to profile S_1, while mistakes on both criteria lead to S_4. Yet, the outcome ensuing from S_1 is further away from the outcome resulting from S_3 than the outcome that would have resulted had more – indeed both – criteria been erroneously measured whereupon S_4 would have emerged. This shows that measurement mistakes do make a difference. It should be emphasized that the violation of proximity preservation occurs in a wide variety of aggregation systems, viz. those that satisfy anonymity and unanimity. This result is not dependent on any particular metric with respect to which the distances between profiles and outcomes are measured. Expressed in another way the result states that in nearly all reasonable aggregation systems it is possible that a small number of measurement errors has greater impact on evaluation outcomes than a big number of errors.

The theorem – when interpreted in the multiple-criterion choice context – does not challenge completeness or transitivity of individual preferences, but calls into question the continuity of preferences, i.e. their representation by smooth utility functions.

5   Invoking Reasons

The upshot of the preceding is that all assumptions underlying the utility maximization theory can be questioned, not only from the descriptive accuracy but also from the normative point of view. The deviations from the assumptions described above are not unreasonable or irrational. In fact, it can be argued that they are just the opposite, viz. based on reasons for having opinions (cf. Dietrich and List 2013). Incompleteness of preference relations as exhibited by Ostrogorski's paradox is a result of a systematic comparison of alternatives using a set of criteria and a set of aspects or dimensions or purposes ("functions") that the alternatives are associated with. There is a reason for the incompleteness: simple majority rule gives different results when row-column aggregation or column-row aggregation is resorted to. The simple majority rule is not the sole culprit: the paradox can occur with super-majority rules as well. The point is that one can build a plausible argument for the incompleteness under some circumstances.

The same goes for intransitivity. The argument is, however, somewhat different in invoking reasons for having a given binary preference: the reason for preferring A to B may differ from the one for putting B ahead of C and this, in turn, may differ from the basis for preferring A to C or vice versa. As May (1954) pointed out, the basic sets from which choices are made are different in each of these three cases.

The eventual failure on continuity rests on yet another consideration. By Baigent's theorem any rule that is anonymous (does not discriminate for or against alternatives) and respects unanimity (in agreeing with the ranking that is identical on all criteria) can lead to discontinuities. One could argue that any reasonable rule is prone to discontinuous utility representations.

To reiterate: the grounds for deviating from the assumptions of utility maximization are normative, not just descriptive. In other words, it makes perfect sense to have preferences that deviate from the assumptions. The question now arises: are there alternatives to these assumptions that could be used in analyzing individual choice behavior? In what follows we shall argue that there are and, moreover, these alternatives provide adequate foundations for institutional design.

6   Dealing with Incomplete, Intransitive and Discontinuous Preferences

The most natural way of handling intransitive preference relations is to start from complete relations and look for methods to aggregate them. This approach has a long history. The most important of the early pioneers is Ernst Zermelo (1929). The starting point is the concept of tournament, i.e. a complete and asymmetric relation. With a finite (and small) number k of alternatives this can conveniently be represented as a $k \times k$ matrix where the element a_{ij} on the ith row and jth column equals 1 whenever ith alternative is preferred to the jth one. Otherwise, the element equals 0.

Given an individual preference tournament we might be interested in forming a ranking that would preserve the essential features of the tournament, while at the same time augmenting it so that a complete and transitive relation emerges. The latter might be necessary e.g. for aggregating individual preference information to end up with a social ranking or choice. By the fundamental result of

Edward Szpilrajn (1930) every partial order – that is a asymmetric and transitive relation – has a linear extension. In other words, if the individual gives a preference relation that is asymmetric (strict preferences only) and transitive, but not complete (not all pairs of alternatives are comparable), then preference rankings can be constructed that preserve those aspects provided by the individual. The problem is that the resulting rankings are rarely unique. In fact, if x and y are two non-comparable alternatives in the relation given by the individual, there are rankings in which $x \succ y$ and rankings in which $y \succ x$ (Dushnik and Miller 1941). Thus, there seems to be no general way of extending a partial order into a unique linear one.

However, it can be argued that tournaments put less structure into individual preferences than partial orders.[5] After all, they are complete and asymmetric, not necessarily transitive. Over past decades many ways of translating tournaments into rankings have been suggested. The usual way – called scoring method by Rubinstein (1980) – is the straight-forward summing of row entries in the tournament matrix whereby one ends up with $s_i = \sum_j a_{ij}$ for each alternative i. The ranking over the alternatives is then determined by the order of scores. The resulting ranking is, of course, weak since several alternatives may receive the same score.

The scoring method may, however, lead to an outcome ranking where a higher rank is given to an alternative that is deemed inferior to one or several of the lower ranked ones. Several methods to avoid this problem has been suggested. Thus, for example, Goddard's (1983) proposal is to choose those rankings that minimize the number of times a binary preference between any two alternatives is upset (i.e. reversed) in the outcome ranking.[6] Upon closer inspection this proposal turns out to be similar to Kemeny's (1959) rule. Viewed as a social choice function this rule has a host of desirable properties (see, e.g. Nurmi 2012, 257). It is, however, intended for finding the "closest" social ranking for any given set of individual rankings over several alternatives. A function that – given a set of individual preference tournaments – looks for the collective one that is closest to the individual tournaments in a specific sense is – regrettably nowadays largely forgotten – Slater's (1961) rule. It seems identical to the rule that Goddard advocates. It works, as was already stated, on the basis of individual tournaments, i.e. complete and asymmetric relations. It then generates all $k!$ complete and transitive relations (strict rankings) that can be obtained from the k alternatives and converts them into tournament matrices. Each of these generated matrices is then a candidate for the collective preference tournament (i.e. the winning tournament). The winning tournament has the distinction that it is closest to the individual tournaments in the sense that it requires the minimum number of changes from 0 to 1 or vice versa in individual opinions to be unanimously adopted.

---

[5] Admittedly, this claim rests on a specific intuitive concept of structure.

[6] Goddard is not the first one to suggest this method. For earlier discussions, see Kendall (1955) and Brunk (1960).

The principle of Slater's rule can, of course, be used in individual decision making as well. To wit, given an individual preference tournament one generates the tournaments corresponding to all $k!$ preference rankings involving the same number of alternatives. One then determines whether the individual tournament coincides with one of them. If it does, then this gives us the ranking we are looking for. Otherwise one determines which of the generated tournaments is closest to the individual's. The closest one indicates the ranking. It may happen that there are several equally close tournaments and thus there may be several "solutions".

Zermelo's (1929) approach to tournaments is based on observations of chess playing contests which often take the form of a tournament.[7] Each player plays against every other player several times. The outcome of each game is either a victory of one player or a tie. We assume that the games are independent binomial trials so that the probability of player i beating player j is p_{ij}. Zermelo then introduces the concept *Spielstärke*, playing strength, denoted by V_i, that determines the winning probability as follows:

$$p_{ij} = \frac{V_i}{V_i + V_j}.$$

The order of the V_i values is the ranking of the players in terms of playing strength. Apparently player i is ranked no lower than player j if and only if $p_{ij} \geq 1/2$, i.e. players with greater strength defeat contestants with smaller strength more often than not. Now, given the matrix A of results, i.e. a $k \times k$ matrix of 0's and 1's denoting losses and victories of the alternatives represented by the rows, Zermelo defines maximum likelihood estimates, denoted by v_i, for the playing strengths of players. Consider any k vector of strengths v. One can associate with it the probability that the observed matrix A is the result of the tournament when the strengths are distributed according to v. The probability is the following:

$$p(v) = \prod_{i,j}\left(\frac{v_i}{v_i + v_j}\right).$$

and this is what is to be maximized. Conditions under which a unique maximizing vector of strengths can be found are discussed by Zermelo and found to be rather general. A particularly noteworthy property of the Zermelo rankings is that they always coincide with the rankings in terms of scores defined above. So, were one interested in rankings only, the easy way to find them is simply to compute the scores. However, the v_i values give us more information about the players than just their order of strength; it also reveals how much stronger player i is when compared with player j.

---

[7] The differences between Zermelo's and Goddard's approaches are cogently analyzed by Stob (1985). Much of what is said in this and the next paragraph is based on Stob's brief note.

Leaving aside now the game context and looking at Zermelo's method from the point of view of fuzzy systems, it is not difficult to envision a new interpretation whereby the outcome matrix expresses the individual's choice between pairs of alternatives. The values V_i and their estimates v_i can be viewed as values of *desirability* of alternatives. A ranking based on desirability of alternatives is certainly a worthy goal of inquiry and Zermelo's approach gives us plausible way to achieve it.[8]

The above remarks pertain to situations where we are given an individual preference tournament and, for one reason or another, are looking for a ranking that would best approximate it. It is, however, quite easy to envision situations where no ranking at all is required, but rather choice of a subset of "best alternatives". These kinds of situations have been dealt with elsewhere (see Aizerman and Aleskerov 1995; Nurmi and Kacprzyk 1991; Kacprzyk and Nurmi 2000; Kacprzyk et al. 2008, 2009).

7 Concluding Remarks

We have attempted to show above that there are quite plausible reasons for individuals to deviate from the behavior dictated by preference-based utility maximization theory. Indeed, behavior based on reasons would seem to be particularly prone to these kinds of deviations. Rankings being the basic concept underlying the maximization theory, our main conclusion is that alternatives to ranking assumption already exist. One of these, individual preference tournament, has been discussed at some length above. Of particular interest is the re-discovery of Zermelo's approach to tournaments since it provides a natural link between directly observable pairwise choices and the underlying fuzzy notion of desirability. It thus provides a method for estimating fuzzy preference degrees for observational data.

References

Aizerman, M., Aleskerov, F.: Theory of Choice. North-Holland, Amsterdam (1995)

Baigent, N.: Preference proximity and anonymous social choice. Q. J. Econ. **102**, 161–169 (1987)

Baigent, N., Eckert, D.: Abstract aggregations and proximity preservation: an impossibility result. Theor. Decis. **56**, 359–366 (2004)

Baigent, N., Klamler, C.: Transitive closure, proximity and intransitivities. Econ. Theor. **23**, 175–181 (2004)

Bar-Hillel, M., Margalit, A.: How vicious are cycles of intransitive choice? Theor. Decis. **24**, 119–145 (1988)

---

[8] We shall here ignore the ties in pairwise comparisons. These can certainly be dealt with in fuzzy systems theory. Also the tournament literature referred to here is capable of handling them. Ties are typically considered as half-victories, i.e. given a value $1/2$ in the tournament matrices.

Brunk, H.: Mathematical models for ranking from paired comparisons. J. Am. Stat. Assoc. **55**, 503–520 (1960)

Daudt, H., Rae, D.: Social contract and the limits of majority rule. In: Birnbaum, P., Lively, J., Parry, G. (eds.) Democracy, Consensus and Social Contract, pp. 335–357. Sage, London and Beverly Hills (1978)

Dietrich, F., List, C.: A reason-based theory of rational choice. Nous **47**, 104–134 (2013)

Dushnik, B., Miller, E.: Partially ordered sets. Am. Math. Mon. **63**, 600–610 (1941)

Eckert, D., Lane, B.: Anonymity, ordinal preference proximity and imposed social choices. Soc. Choice Welf. **19**, 681–684 (2002)

Fishburn, P.C.: The irrationality of transitivity in social choice. Behav. Sci. **15**, 119–123 (1970)

Goddard, S.: Ranking in tournaments and group decision making. Manag. Sci. **29**, 1384–1392 (1983)

Harsanyi, J.C.: Rational Behavior and Bargaining Equilibrium in Games and Social Situations. Cambridge University Press, Cambridge (1977)

Kacprzyk, J., Zadrożny, S., Fedrizzi, M., Nurmi, H.: On group decision making, consensus reaching, voting and voting paradoxes under fuzzy preferences and a fuzzy majority: a survey and some perspectives. In: Bustince, H., Herrera, F., Montero, J. (eds.) Fuzzy Sets and Their Extensions: Representation, Aggregation and Models. Studies in Fuzziness and Soft Computing, vol. 220, pp. 263–395. Springer, Heidelberg (2008). https://doi.org/10.1007/978-3-540-73723-0_14

Kacprzyk, J., Zadrożny, S., Nurmi, H., Fedrizzi, M.: Fuzzy preferences as a convenient tool in group decision making and a remedy for voting paradoxes. In: Seising, R. (ed.) Views on Fuzzy Sets and Systems from Different Perspectives. Studies in Fuzziness and Soft Computing, vol. 243, pp. 345–360. Springer, Heidelberg (2009). https://doi.org/10.1007/978-3-540-93802-6_16

Kacprzyk, J., Nurmi, H.: Social choice and fuzziness: a perspective. In: Fodor, J., De Baets, B., Perny, P. (eds.) Preferences and Decisions under Incomplete Knowledge. Physica-Verlag, Heidelberg (2000)

Kemeny, J.: Mathematics without numbers. Daedalus **88**, 571–591 (1959)

Kendall, M.: Further contributions to the theory of paired comparisons. Biometrics **11**, 43–62 (1955)

May, K.O.: Intransitivity, utility and aggregation of preference patterns. Econometrica **22**, 1–13 (1954)

Nurmi, H.: On the relevance of theoretical results to voting system choice. In: Felsenthal, D., Machover, M. (eds.) Electoral Systems. Studies in Choice and Welfare, pp. 255–274. Springer, Heidelberg (2012). https://doi.org/10.1007/978-3-642-20441-8_10

Nurmi, H., Kacrpzyk, J.: On fuzzy tournaments and their solution concepts in group decision making. Eur. J. Oper. Res. **51**(1991), 223–232 (1991)

Rubinstein, A.: Ranking the participants in a tournament. SIAM J. Appl. Math. **38**, 108–111 (1980)

Savage, L.: The Foundations of Statistics. Wiley, New York (1954)

Slater, P.: Inconsistencies in a schedule of paired comparisons. Biometrica **48**, 303–312 (1961)

Stob, M.: Rankings from round-robin tournaments. Manag. Sci. **31**, 1191–1195 (1985)

Szpilrajn, E.: Sur l'extension de l'ordre partiel. Fundam. Math. **16**, 386–389 (1930)

Zermelo, E.: Die Berechnung der Turnier-Ergebnisse als ein Maximumproblem der Wahrscheinlichkeitsrechnung. Math. Zeit. **29**, 436–460 (1929)

A Note on Positions and Power of Players in Multicameral Voting Games

Marcin Malawski[✉]

Leon Koźmiński University, Jagiellońska 59, 03-301 Warszawa, Poland
mmn@kozminski.edu.pl

Abstract. A multicameral simple game is an intersection of a number of simple games played by the same set of players: a coalition is winning in the multicameral game if and only if it is winning in all the individual games played. Examples include decision rules in multicameral parliaments where a bill must be passed in all the houses of the parliament, and voting rules in the European Union Council where a winning coalition of countries must satisfy two or three independent criteria. This paper is a preliminary study of relations between the positions and power indices of players in the "chamber" games and in the multicameral game obtained as the intersection. We demonstrate that for any power index satisfying a number of standard properties, the index of a player in the multicameral game can be smaller (or greater) than in all the chamber games; this can occur even when the players are ordered the same way by desirability relations in all the chamber games. We also observe some counterintuitive effects when comparing the positions and decisiveness of players. However, as expected, introducing an additional chamber with all the players equal (a one man - one vote majority game) to a complete simple game reduces all the differences between the Shapley-Shubik indices of players.

Keywords: Simple games · Multicameral voting · Complete games · Power indices · Reducing power inequalities

1 Introduction and Prerequisites

Multicameral voting systems are ubiquitous in contemporary politics. On the one hand, multicameral (usually bicameral) voting is almost standard in parliaments. Voters are formally individual MPs. However, because of the existence of political parties and of party discipline during votes, it is usual to treat *parties*, each having a number of representatives (= votes) in each chamber of the parliament, as voters (players)[1]; a bill is passed if and when it is accepted by the required majorities of votes in each chamber. In game theoretic language, the same players – i.e., parties – play different weighted voting games in different houses of the parliament and the final outcome of the game is determined by the outcomes

---

[1] Of course, if some independent MPs are present, they also are voters.

© Springer International Publishing AG 2017
N.T. Nguyen et al. (Eds.): TCCI XXVII, LNCS 10480, pp. 40–52, 2017.
https://doi.org/10.1007/978-3-319-70647-4_3

of all these voting games. On the other hand, in recent decades multicameral voting has appeared in a different setting in which the "chambers" and the representatives of players are virtual but the rules determining the group decision are exactly the same and thus can be described by multicameral voting as well. This is the case of voting systems in the European Union Council (formerly, the Council of Ministers). Both under the Treaty of Nice and under the Lisbon Treaty currently in force, a group of countries must fulfill a number of independent criteria to enforce a joint decision. First, it must constitute a (qualified) majority of the member countries, and second, it must represent a (qualified) majority of the EU population. Additionally, under the Nice Treaty a third weighted majority voting game was designed by assigning "political" weights to each state and a threshold necessary for a coalition to become winning. Thus, we have had a tricameral voting system and now we have a bicameral one[2]. In one chamber each player – country – has one vote, in another the number of votes a country possesses is proportional to its population, and in the third one, according to the Treaty of Nice, the weights of each country were stipulated. Of course, no voting in particular houses takes place, but if it comes to voting by the countries, the decision is reached *as if* it took place.

Notice that, in contrast to parliaments, the EU voting rules did not evolve naturally but have been designed and approved with the explicit purpose of assigning to each country an amount of "power" acceptable to all EU member states. This raises a question of whether, and how, adding another chamber with specific weights of voters to an existing voting system can influence the voters' positions in the system and, more specifically, their power. This paper collects some observations and statements on this subject.

It is standard today to analyse voting power in the terminology of cooperative games, and such an approach will also be used in this paper. In particular, we shall speak of players instead of voters, having in mind that players are parties or countries. Thus, let $N = \{1, 2, \ldots, n\}$ be a finite set of players and let v be the <u>characteristic function</u> – any real function defined on the set of all <u>coalitions</u>, i.e. all subsets of N, such that $v(\emptyset) = 0$. The pair (N, v) is an n-person cooperative game. Since n will normally be fixed in this paper, we shall identify any cooperative game by its characteristic function, and usually write simply v instead of (N, v).

An important subclass of cooperative games are weighted majority voting games (WMV games). An n-person WMV game is described by a system $[\mu; \lambda_1, \lambda_2, \ldots, \lambda_n]$ of $n + 1$ real numbers such that $\forall i \; \lambda_i > 0$ and $0 < \mu \leq \Lambda_N = \sum_{j=1}^{n} \lambda_j$. The components λ_i are <u>weights</u> of the players, and μ is the required

---

[2] Actually, the Lisbon voting system slightly differs from the bicameral one described above because of an additional clause requiring every *blocking* coalition to consist of at least four member states – i.e., making each group of at least 25 states (in the EU-28) winning regardless of its population share. However, this modification is indeed minor; in the EU-28, it only affects the status of 9 out of 3276 such groups of countries.

majority (quota); it is often assumed (in accordance with the term "majority") that $\mu > \frac{\lambda_N}{2}$. The system $[\mu; \lambda_1, \lambda_2, \ldots, \lambda_n]$ defines a weighted majority voting game (N, v) with the characteristic function given by

$$v(S) = \begin{cases} 1 & \text{when } \sum_{j \in S} \lambda_j \geq \mu \\ 0 & \text{when } \sum_{j \in S} \lambda_j < \mu \end{cases}.$$

We shall also denote this game by $[\mu; \lambda_1, \lambda_2, \ldots, \lambda_n]$. Clearly, many different weights/quota systems define the same game (i.e. with the same characteristic function); also, one can always select a system where all the weights and quota are integers.

In particular, in a one man - one vote majority voting game all the players' weights are equal to one and so $v(S) = 1$ if and only if S consists of no less than μ players. The n-person one man - one vote MVG with quota μ will be denoted by $m_{n,\mu}$.

Majority voting games, on the other hand, belong to a broader important subclass of cooperative games – simple games. A game (N, v) is a simple game if its characteristic function satisfies

1. for every coalition $S \subseteq N$ $v(S) = 0$ or $v(S) = 1$,
2. $v(N) = 1$,
3. if $S \subseteq T$, then $v(S) \leq v(T)$ (monotonicity).

Thus, in a simple game w, winning coalitions are those belonging to $w^{-1}(1)$, and losing coalitions – those in $w^{-1}(0)$. A coalition is minimal winning in w if it is winning in w but all its proper subsets are losing. For a given simple game w, let us denote the sets of winning, losing and minimal winning coalitions in w by $\mathcal{W}(w)$, $\mathcal{L}(w)$ and $\mathcal{MW}(w)$, respectively. It is obvious that any one of the sets $\mathcal{W}(v)$, $\mathcal{L}(v)$ uniquely determines the game v, and it is easily proved that the same is true for the set $\mathcal{MW}(v)$. We shall denote by $\mathcal{P}_n$ the set of all n-person simple games, and by $\mathcal{P}^* = \bigcup_{n=1}^{\infty} \mathcal{P}_n$ the set of all simple games with a finite number of players.

It is obvious that there are simple games which are not (weighted) majority voting games, i.e., their characteristic function does not admit representation by any weights system λ and quota μ. The reason is that a weights system naturally orders the players according to their "strength": the greater a player's weight, the more he can contribute to coalitions. In practice, however, it can happen that one coalition finds it more worthwhile to incorporate player i than player j but another coalition's preference is the opposite. For instance, in a parliament consisting of two chambers a party (or coalition) having a majority in the upper house but short of a majority in the lower house by just one vote finds an additional independent MP in the lower house useful but one in the upper house useless, while a coalition with a majority in the lower house but not in the upper house will see it exactly the opposite way. We thus see that a "bicameral" simple game sometimes cannot be described as a WMV game.

Formally, a simple game (N, v) is a <u>multicameral</u> voting game if it is the intersection (minimum) of at least two weighted majority voting games with the same sets of players. An <u>intersection</u> of two simple games (N, y) and (N, z) is another simple game $(N, y \wedge z)$ defined by

$$(y \wedge z)(S) = y(S) \cdot z(S) = \min(y(S), z(S)) \quad \forall S \in \mathcal{N},$$

and a simple game (N, v) is <u>k-cameral</u> if it is of the form

$$v = v_1 \wedge v_2 \wedge \ldots \wedge v_k$$

where $v_j = [\mu_j; \lambda_{1,j}, \lambda_{2,j}, \ldots, \lambda_{n,j}]$ for $j = 1, 2, \ldots k$, that is, all the games $v_1, v_2, \ldots, v_k$ are weighted majority voting games. The interpretation is obvious: a coalition S is winning in the game v if and only if it has the required majority in each of the WMV games $v_1, v_2, \ldots, v_k$. These games can be viewed as separate chambers in which each player $1, 2, \ldots, n$ has some representation. (Obviously, this representation can be null – this is the case for an independent MP in one of the chambers or for a party which has no MPs in some chamber).

It is important to stress that this notion of multicamerality differs from that studied by Felsenthal, Machover and Zwicker [3] who use this term for the situation in which the sets of players in each of the "houses" (weighted voting games) are disjoint. Their notion can be interpreted as a very special case of each player (party, country, ...) having representatives only in one chamber and, needless to say, is of very limited practical importance. It would describe, for instance, voting in the EU under a rule requiring both the majority of "old" members and the majority of "new" members to vote for a motion for it to be accepted by the Union.

A well-known theorem (see e.g. Taylor and Zwicker [6]) says that every simple game is a multicameral voting game. However, the practical use of this result is also limited because the proof requires a large number k of "chambers" – for an arbitrary simple game v, it can be as large as $\#\mathcal{MW}(v)$, much larger than the number of players. A non-trivial open problem is finding the *minimal* number of WMV games in the multicameral representation of any given simple game.

2   Desirability and Positions of Players in Multicameral Games

In this section we discuss the relations between players' rôles, or positions, in the games played in individual chambers and in the multicameral game obtained as the intersection of these games. We shall sometimes use the term "chamber games" for the games $v_1, v_2, \ldots, v_k$ whose intersection $v = v_1 \wedge v_2 \wedge \ldots \wedge v_k$ will be discussed.

Let us first consider minimal winning coalitions.

Observation 1. Let $v = v_1 \wedge v_2 \wedge \ldots \wedge v_k$ where $v_1, v_2, \ldots, v_k$ are any simple games on the same set of players (not necessarily WMV games). Then

1.

$$MW(v) \supseteq \bigcup_{i=1}^{k} \left(MW(v_i) \cap \bigcap_{j \neq i} W(v_j) \right)$$

2. The inclusion can be strict: $MW(v) \setminus \bigcup_{i=1}^{k} MW(v_i) \neq \emptyset$
3. If $MW(v) \setminus \bigcup_{i=1}^{k} MW(v_i) \neq \emptyset$ and $v_1, v_2, \ldots, v_k$ are WMV games, then there exist two players i , j such that in some chamber games v_l , v_m $(1 \leq l, m \leq k)$ $\lambda_{i,l} > \lambda_{j,l}$ and $\lambda_{i,m} < \lambda_{j,m}$.

Proof of 3. Let the coalition T be minimal winning in the game $v_1 \wedge v_2$ but neither in v_1 nor in v_2. However, T must be winning in both v_1 and v_2. This means that there exist players $i, j \in T$ such that $T \setminus \{i\}$ is winning in v_1 and $T \setminus \{j\}$ in v_2. These players must be distinct since otherwise $T \setminus \{i\} = T \setminus \{j\}$ would be winning in $v_1 \wedge v_2$, in contradiction with the assumption that T is minimal winning. Thus, $T \setminus \{i\}$ is losing in v_2 and $T \setminus \{j\}$ in v_1, which in the case of WMV games implies that $\lambda_{i,1} < \lambda_{j,1}$ and $\lambda_{i,2} > \lambda_{j,2}$.

Example 1. Take $n = 5$ and $v_1 = [12; 5, 5, 2, 2, 2]$, $v_2 = [12; 2, 2, 5, 5, 2]$. Then $v = v_1 \wedge v_2 = [11; 3, 3, 3, 3, 1]$ and so there is only one coalition in $MW(v)$, namely $S = \{1, 2, 3, 4\}$, but it is neither minimal winning in v_1 nor in v_2.

The position of a player in a simple game is best described by his decisiveness. Player i is <u>decisive in coalition T</u> in the simple game w if and only if $w(T) = 1$ and $w(T \setminus \{i\}) = 0$, i.e. if the winning coalition T becomes losing when i leaves it. For a weighted majority voting game $w = [\mu; \lambda_1, \lambda_2, \ldots, \lambda_n]$ this is equivalent to two inequalities $\lambda_T = \sum_{j \in T} \lambda_j \geq \mu$ and $\lambda_T - \lambda_i < \mu$. We shall denote by $D(i, w)$ the set of all coalitions in which i is decisive (in the game w). Intuitively, the larger the set $D(i, w)$, the stronger is the position of player i in w.

The following simple yet useful result shows how a player's decisiveness in the chamber games translates to decisiveness in the bicameral game.

Proposition 1. *When v_1, v_2 are two simple games and $v = v_1 \wedge v_2$, then for every player i*

(a) $D(i, v_1) \cap D(i, v_2) \subseteq D(i, v) \subseteq D(i, v_1) \cup D(i, v_2)$,
(b). *both inclusions in (a) can be strict*,
(c) $D(i, v) = (D(i, v_1) \cap W(v_2)) \cup (W(v_1) \cap D(i, v_2))$.

The proofs are quite straightforward and are omitted.

A weakest possible position of a player is that of a <u>null player</u> who is not decisive in any coalition, and the strongest one is being a <u>dictator</u> who is decisive in all the coalitions to which he belongs. Formally, in a simple game w,

player i is a null player $\Leftrightarrow D(i, w) = \emptyset \Leftrightarrow \forall_{T \subseteq N} w(T) = w(T \cup \{i\})$,

player i is a dictator $\Leftrightarrow \forall_{T \subseteq N} (w(T) = 1 \Leftrightarrow i \in T)$
    $\Leftrightarrow w(\{i\}) = 1$ and $w(N \setminus \{i\}) = 0$.

Games with dictators are of relatively little interest because if there is a dictator in a simple game, then all the other players are null players and the game is uniquely determined. A slightly weaker position than the dictator's is that of a veto player who must belong to all winning coalitions (and thus is decisive in each of them). In a simple game w,

$$\text{player } i \text{ is a veto player } \Leftrightarrow \forall_{T \subseteq N} \, (w(T) = 1 \Rightarrow i \in T) \Leftrightarrow i \in \bigcap \mathcal{W}(v).$$

The following observations summarize the relations between dictators, veto players and null players in chamber games and in their intersection.

Observation 2. Let $v = v_1 \wedge v_2 \wedge \ldots \wedge v_k$ where $v_1, v_2, \ldots, v_k$ are any simple games (again, not necessarily WMV games). Then

1. If player i is a dictator in some of the games $v_1, \ldots, v_k$, then i is a veto player in v.
2. If i is dictator in v, then i is also a dictator in some[3] of the games $v_1, \ldots, v_k$.

The proofs are obvious, using the definition of $y \wedge z$ and the characterization of a dictator player by $(w(\{i\}) = 1$ and $w(N \setminus \{i\}) = 0)$.

Observation 3. Let $v_1, v_2, \ldots, v_k$ and v be as in Observation 2. Then

1. Player i is a veto player in v $\Leftrightarrow$ i is a veto player in at least one of the games $v_1, \ldots, v_k$.
2. If i is a null player in all the games $v_1, v_2, \ldots v_k$, then i is a null player in v.
3. A null player in v may be non-null in all the games $v_1, v_2, \ldots v_k$.

Proof of 1. and 2. To prove the "only if" implication in 1., assume that i is not a veto player in any of the games $v_1, \ldots, v_k$. Then, for each game v_j, there exists a coalition $S_j \in \mathcal{W}(v_j)$ such that $i \notin S_j$. The coalition $S = S_1 \cup S_2 \cup \ldots \cup S_k$ is winning in v and does not contain player i, so i is not a veto player in v. The "if" part of 1. and 2. are straightforward corollaries from the definition of $y \wedge z$.

Example 2. In the games v_1, v_2 and v of Example 1, player 5 is non-null in both v_1 and v_2 (being decisive in the 3-person coalition with two largest players) but is (the only) null player in v.

The positions of all the other players who are neither null nor veto players are in-between – they are all "stronger" (more decisive) that any null player and "weaker" than any veto player. Among themselves, they also can be often compared according to their decisiveness. We shall say that in a simple game (N, w) player i is

---

[3] In general, it is not true that a dictator in v must be a dictator in every chamber game: to see it, take v_1 – the game in which player 1 is a dictator and v_2 – the game in which winning coalitions are exactly those containing player 1 or players 2 and 3. Then $v = v_1 \wedge v_2 = v_1$ but clearly 1 is not a dictator in v_2. However, the above stronger implication holds when all the chamber games are assumed to be superadditive, i.e. do not contain two disjoint winning coalitions.

<u>not weaker</u> than player $j \Leftrightarrow \forall_{T \not\ni i,j} \; w(T \cup i) \geq w(T \cup j)$;
<u>stronger</u> than player $j \Leftrightarrow \forall_{T \not\ni i,j} \; w(T \cup i) \geq w(T \cup j)$
        and $\exists_{T \not\ni: i,j} : w(T \cup i) > w(T \cup j)$.

This will be denoted by $i \succeq_w j$ (i not weaker than j in w) and $i \succ_w j$ (i stronger than j in w). The relation $\succeq_w$ is the <u>desirability</u> relation introduced and studied by Isbell [4]; the relation $\succ_w$ is the <u>strict part</u> of $\succeq_w$.

A simple game w is <u>complete</u> (Carreras and Freixas [2]) if its desirability relation is complete, i.e. if for each pair of players $i, j \in N$ at least one of $i \succeq_w j$ and $j \succeq_w i$ is true. Obviously, all weighted majority voting games are complete, but there are complete simple games which are not WMV games, and this is precisely the effect of "multicamerality". For instance, both the Nice game and the Lisbon game in the EU Council are complete but cannot be represented as single WMV games.

Concerning the connections between desirability relations in chamber games and in their intersection, it turns out that being "not weaker" is inherited from chamber games to the bicameral game, but being "stronger" is not.

Observation 4. Let v_1, v_2 be two simple games and let $v = v_1 \wedge v_2$. Then

1. If $i \succeq_{v_1} j$ and $i \succeq_{v_2} j$, then also $i \succeq_v j$.
2. It is not true that $((i \succ_{v_1} j \text{ and } i \succ_{v_2} j) \Rightarrow i \succ_v j))$.

Proof. To prove 1., assume the contrary: let $i \succeq_{v_1} j$ and $i \succeq_{v_2} j$ but not $i \succeq_v j$. Then there must exist a coalition $T \subset N$ such that $i, j \notin T$, $v(T \cup i) = 0$ and $v(T \cup j) = 1$. This means that the coalition $T \cup j$ is winning in both chamber games v_1 and v_2, and $T \cup i$ is losing in at least one of these games, which contradicts $i \succeq_{v_1} j$ or $i \succeq_{v_2} j$. 2. is demonstrated by the following counterexample, obtained as a slight modification of Example 1 by adding a null player 6 in both games v_1 and v_2.

Example 3. Take $n = 6$, $v_1 = [16 \, ; 7, 7, 2, 2, 2, 1]$, $v_2 = [16 \, ; 2, 2, 7, 7, 2, 1]$.
In both these games player 6 is null and player 5 is non-null, so $5 \succ_{v_1} 6$ and $5 \succ_{v_2} 6$. But both 5 and 6 are null players in $v = v_1 \wedge v_2 = [34 \, ; 9, 9, 9, 9, 4, 2]$ and so it is not true that $5 \succ_v 6$.

Remark 1. We have observed a number of "paradoxes" in multicameral games: minimal winning coalitions in the multicameral game which are not minimal winning in any of the chamber games, players that are non-null in all the chamber games but null in the multicameral game, and two players being of equal "strength" in the bicameral game despite one being stronger than the other in both chamber games. On the other hand, all the examples in this section relied on incompatible orderings of players in the chambers – one player having greater weight than the other in one house but smaller in another house. It is therefore of interest to check whether the above "paradoxes" indeed result from incompatible orderings.

We shall say that two simple games (N, v_1) and (N, v_2) have <u>incompatible orderings</u> of players if there exists a pair of players $i, j \in N$ such that

$$i \succ_{v_1} j \text{ and } j \succ_{v_2} i.$$

In weighted voting games this amounts to the following (essential) inequalities regarding weights:

$$\lambda_{i,1} > \lambda_{j,1} \text{ and } \lambda_{i,2} < \lambda_{j,2}.$$

If two complete simple games v_1 and v_2 do not have incompatible orderings of players, we say that they have <u>compatible orderings</u>[4]. Notice that the compatibility of orderings does not imply perfect concordance of desirability relations: for instance, the equal desirability of all players $(\forall i, j \, (i \neq j \Rightarrow i \succeq_w j))$ is compatible with any complete ordering of players.

Proposition 2. *Let $v = v_1 \wedge v_2$ with v_1 and v_2 being two complete simple games with compatible orderings of players. Then every minimal winning coalition in the bicameral game v is minimal winning in at least one of the games v_1, v_2.*

Proof. It was shown in the proof of Observation 1, part 3. that if the set $\mathcal{MW}(v) \setminus (\mathcal{MW}(v_1) \cup \mathcal{MW}(v_2))$ is nonempty, then for every coalition T in this set there exist two distinct players $i, j \in T$ such that $T \setminus i$ is winning in v_1 and losing in v_2, and $T \setminus j$ is winning in v_2 and losing in v_1. This means that neither $j \succeq_{v_1} i$ nor $i \succeq_{v_2} j$. But if the games v_1 and v_2 are complete, then this is equivalent to $i \succ_{v_1} j$ and $j \succ_{v_2} i$, implying incompatible orderings of players in v_1 and v_2.

However, other paradoxes of players' positions and "strength" can occur even when the chamber games are complete and the orderings of players are compatible. Examples 4 and 5 below demonstrate this for an even more restrictive case when the chamber games are weighted majority voting games with the same orderings of players' weights.

Example 4. Take $n = 4$, $v_1 = [8 \, ; 7, 4, 2, 1]$ and $v_2 = [6 \, ; 4, 3, 2, 1]$. Player 4 is non-null both in v_1 (being decisive in coalition with player 1) and in v_2 (decisive in coalition with players 2 and 3). However, in the game $v_1 \wedge v_2 = [8 \, ; 5, 3, 3, 1]$ player 4 does not belong to any minimal winning coalition $(\mathcal{MW}(v_1 \wedge v_2) = \{12, 13\})$ and so is a null player. Notice that the desirability relations in both chamber games are not just compatible but perfectly concordant: $1 \succ 2 \approx 3 \succ 4$.

Example 5. $n = 6$, $v_1 = [29 \, ; 20, 8, 6, 5, 4, 2]$, $v_2 = [20 \, ; 9, 7, 6, 5, 2, 1]$.
Clearly, the orderings of players in both chamber games are compatible as results from the orderings of their weights. Moreover, in both games player 5 is stronger than player 6:

$$D(5, v_1) = \{125, 135, 145\}, D(6, v_1) = \{126\} \text{ and}$$
$$D(5, v_2) = \{2345, 23456\}, D(6, v_2) = \emptyset.$$

---

[4] If v_1 and v_2 are not complete, the lack of incompatibilities as defined above need not have any meaningful consequences because the players can be incomparable.

However, in the bicameral game v we have $6 \succeq_v 5$ because both 5 and 6 are null players. This game is given by $\mathcal{MW}(v) = \{123, 124, 134\}$ so it can be represented e.g. as $v = [15 \; ; \; 8, 4, 4, 4, 1, 1]$.

We thus see that, even if players' rôles and desirability in chamber games are compatible, these rôles and desirability can behave in a different way in the multicameral game. Even when the chamber games are WMV games with the same orderings of numbers of seats, a player who is non-null in all the chambers can become null in the multicameral game, and a player i with less seats in all the chambers than another player j – indeed, weaker than j in all the chambers – can become as strong as j in the multicameral game (although reverting the ordering is not possible, as noted in Observation 4.1.). This suggests that numerical measures of players' power in the multicameral game can also differ from their counterparts in the chamber games. We explore this topic in the next section.

3   Comparing Power Indices

Power indices are numerical measures of players' "strength" in simple games, widely applied in political science and in analyses of group decisions. Usually, though not always, they are derived somehow from the players' decisiveness. In this section we show that multicameral voting also leads to some counterintuitive effects in the behaviour of power indices.

Formally, a <u>power index</u> is any function $p : \mathcal{P}^* \rightarrow \bigcup\limits_{n=1}^{\infty} \Delta_n$ such that $\forall n \; p(\mathcal{P}_n) \subset \Delta_n$ (where $\Delta_n = \{x = (x_1, \ldots, x_n) : x_1, \ldots, x_n \geq 0, \sum_{j=1}^{n} x_j = 1\}$ is the n-dimensional probabilistic simplex), with the following <u>null player property</u>:

$$(\forall U \; w(U \cup i) = w(U)) \Rightarrow p_i(w) = 0 .$$

Thus, for every n, p assigns to each n-person simple game (N, v) a probabilistic n-vector $p(v) = (p_1(v), \ldots p_n(v))$ such that for every null player i in v, $p_i(v) = 0$. The components $p_j(v)$ are the <u>power indices of the players</u> and are intended to measure their power in the game v. While assuming the null player property is standard (a null player has no power), imposing normalization – i.e. the requirement that for every simple game (N, v) $\sum_{j=1}^{n} p_j(v) = 1$ – is also common but not universal in the literature since some authors also apply "absolute" (= non-normalized) power indices.

Some other reasonable properties intuitively expected from a power index are:

<u>Non-null player property</u>: $D(i, w) \neq \emptyset \Rightarrow p_i(w) > 0$.
<u>Equal treatment property</u>: $(i \succeq_w j \text{ and } j \succeq_w i) \Rightarrow p_i(w) = p_j(w)$.
<u>Local monotonicity</u>: $i \succeq_w j \Rightarrow p_i(w) \geq p_j(w)$.
<u>Local strict monotonicity</u>: $i \succ_w j \Rightarrow p_i(w) > p_j(w)$.

All these properties point at players' decisiveness as a basis for defining power indices, and indeed most classical indices use it in some way. This holds, in particular, for the two by far most commonly used indices – the Shapley-Shubik index and the Banzhaf index.

The <u>Shapley-Shubik index</u> (Shapley value) [5] ϕ is defined by the formula

$$\phi_i(N, w) = \sum_{T \in D(i,w)} \frac{n!}{(t-1)!(n-t)!}$$

where $t = \#T$. The <u>Banzhaf index</u> [1] b is given by $b_i(N, w) = \dfrac{d(i, w)}{\sum_{j=1}^{n} d(j, w)}$,

where $d(i, w) = \#D(i, w)$ is the number of player i's swings in the game w.

It is well-known and straightforward to check that both these indices are locally monotonic and locally strictly monotonic, and have the non-null player and equal treatment properties.

What are the relations between the power indices of players in the chamber games and in the resulting multicameral voting game? The observations made in Sect. 2 suggest that some negative results should be expected. Indeed, it turns out that the indices of a player in the chamber games do not provide bounds for his index in the multicameral game; moreover, this holds for *any* index satisfying some natural conditions.

Proposition 3. *No power index p having the non-null player property fulfils the inequality*

$$\forall_{v_1, v_2 \in \mathcal{P}_n} \forall_{i=1,2,\ldots,n} \quad p_i(v_1 \wedge v_2) \geq \min\left(p_i(v_1), p_i(v_2)\right).$$

This is a simple corollary from Observation 3.3. Moreover, Example 4 in the preceding section demonstrates that Proposition 3 also remains true when strenghtened by replacing $\mathcal{P}_n$ in its formulation by the smaller class of weighted majority voting games with the same desirability relation among players.

Proposition 4. *No locally strictly monotonic power index p with the equal treatment property fulfils the inequality*

$$\forall_{v_1, v_2 \in \mathcal{P}_n} \forall_{i=1,2,\ldots,n} \quad p_i(v_1 \wedge v_2) \leq \max\left(p_i(v_1), p_i(v_2)\right).$$

The proof is by means of an example.

Example 6. $n = 6$, $v_1 = [19; 5, 5, 5, 5, 3, 3]$, $v_2 = [19; 5, 5, 5, 3, 5, 3]$, $v = v_1 \wedge v_2$. In the game v the minimal winning coalitions are exactly all the 5-person coalitions, that is, $v = m_{6,5} = [5; 1, 1, 1, 1, 1, 1]$. From the equal treatment property, $p(v) = \left(\frac{1}{6}, \frac{1}{6}, \ldots \frac{1}{6}\right)$. On the other hand, local strict monotonicity together with the equal treatment property imply

$$p_1(v_1) = \ldots = p_4(v_1) > p_5(v_1) = p_6(v_1), \text{ and}$$
$$p_1(v_2) = p_2(v_2) = p_3(v_2) = p_5(v_2) > p_4(v_2) = p_6(v_2),$$

so, since p is a normalized index, $p_1(v_1) > \frac{1}{6}$ and $p_1(v_2) > \frac{1}{6}$, and $p_6(v_1) < \frac{1}{6}$ and $p_6(v_2) < \frac{1}{6}$. Thus, $p_6(v_1 \wedge v_2) > \max\left(p_6(v_1), p_6(v_2)\right)$.

Notice that this example again relies on incompatible orderings of players in v_1 and v_2. We currently do not know whether it can be improved to involve chamber games with compatible orderings violating the inequality in Proposition 4.

Corollary 1. *A player's power in a weighted multicameral voting game, as measured by the most canonical indices like the Banzhaf or Shapley-Shubik index, can be greater (or smaller) than the same player's power in each of the chambers.*

This means, in particular, that some anomalies are possible when multicameral simple games are designed for the purpose of making group decisions. A player can become worse off (or better off) in terms of a power measure in the multicameral game than he was in each of the chamber WMV games; the first possibility is even present when the chamber games have compatible orderings. This results from the behaviour of decisiveness and desirability, and in general occurs independently of the power index used. The architects of multicameral voting systems should take all this into account and carefully compute the players' power in their products.

4   Reducing Power Inequalities by Adding a One Man - One Vote Game

In this section we consider a particular case of a multicameral game in which in one of the chambers a one man - one vote majority game is played. Since in this chamber all players are equal, its existence could be expected to mitigate power inequalities between players arising in other chamber(s)[5]. We show that this is true at least when the game played outside this "egalitarian" chamber is complete and the Shapley-Shubik index is applied to measure the power of players.

For any simple game (N, w) and any natural number $k \leq n$ we shall denote

$$w|_k = w \wedge m_{n,k}$$

where $m_{n,k}$ is the n-person one man - one vote voting game with quota k. Thus, $w|_k$ is derived from w by adding an extra criterion that every winning coalition must contain at least k players.

Observation 5. The following properties of the games $w|_k$ are easy to verify:

1. if $k \leq c(w)$, where $c(w)$ is the cardinality of the smallest winning coalition in w, then $w|_k = w$,

---

[5] In particular, we presume that mitigating inequalities (together with pleasing small countries) was precisely the purpose of including one country - one vote games in both the Nice and Lisbon treaties. Given Theorem 1 in this section and the fact that the games played in other "chambers" under voting rules in the EU Council are complete, this operation definitely achieved its aim.

2. $w|_n = m_{n,n}$ – the unanimity game,
3. $T \in D(i, w|_k) \Leftrightarrow i \in T, T \in \mathcal{W}(w)$ and either $(T \in D(i, w)$ and $\#T > k)$ or $\#T = k$.

Theorem 1. *Let (N, v) be any complete simple game, and let k, l be any two integers such that $1 \leq k < l \leq n$. Then for every pair of players $i, j \in N$*

$$|\phi_j(v|_l) - \phi_i(v|_l)| \leq |\phi_j(v|_k) - \phi_i(v|_k)| \leq |\phi_j(v) - \phi_i(v)|.$$

Proof. It is sufficient to prove that for pairs of neighbouring integers $k, k + 1$

$$|\phi_j(v|_{k+1}) - \phi_i(v|_{k+1})| \leq |\phi_j(v|_k) - \phi_i(v|_k)| \tag{1}$$

and then to apply the obtained inequality repeatedly.

Let us fix two players i and j in v and assume without loss of generality that $i \succ_v j$ (if neither $i \succ_v j$ nor $j \succ_v i$, then $i \approx_v j$ because v is a complete game, and all inequalities trivially reduce to equalities).

For any integer $t \leq n$ denote: $R_t = \frac{n!}{(t-1)!(n-t)!}$. In this notation, the Shapley-Shubik index takes the form $\phi_i(N, v) = \sum_{T \in D(i,v)} R_t$ where $t = \#T$. For brevity, we assume that all the sums below are taken only over coalitions including player i. Applying Observation 5.3., we have for any player i

$$\phi_i(v|_k) - \phi_i(v|_{k+1}) =$$

$$= \sum_{T: t=k, T \in \mathcal{W}(v)} R_t + \sum_{T: t>k, T \in D(i,v)} R_t - \left(\sum_{T: t=k+1, T \in \mathcal{W}(v)} R_t + \sum_{T: t>k+1, T \in D(i,v)} R_t \right)$$

$$= \sum_{T: t=k, T \in \mathcal{W}(v)} R_k + \sum_{T: t=k+1, T \in D(i,v)} R_{k+1} - \sum_{T: t=k+1, T \in \mathcal{W}(v)} R_{k+1}$$

$$= R_k \, \omega_{i,k} - R_{k+1} \, \nu_{i,k+1}$$

where

$$\omega_{i,k} = \#\{U : i \in U, \#U = k \text{ and } U \in \mathcal{W}(v)\}$$

and

$$\nu_{i,k+1} = \#\{U : i \in U, \#U = k + 1 \text{ and } U \in \mathcal{W}(v) \setminus D(i, v)\}.$$

Similarly, $\phi_j(v|_k) - \phi_j(v|_{k+1}) = R_k \omega_{j,k} - R_{k+1} \nu_{j,k+1}$. Now it is not difficult to see that if $i \succ_v j$, then for any integer $k \leq n$

$$\omega_{i,k} \geq \omega_{j,k} \text{ and } \nu_{i,k} \leq \nu_{j,k}$$

– a stronger player belongs to at least as many winning coalitions of a given size as a weaker one, and to at most as many winning coalitions of a given size in which he is not decisive as a weaker one (these remain winning also if he leaves). Thus, $\phi_i(v|_k) - \phi_i(v|_{k+1}) \geq \phi_j(v|_k) - \phi_j(v|_{k+1})$ and so

$$\phi_i(v|_k) - \phi_j(v|_k) \geq \phi_i(v|_{k+1}) - \phi_j(v|_{k+1}).$$

Both differences in the above must be nonnegative since $i \succeq_{v|_k} j$ in all games $v|_k$. This proves (1). Moreover, $\phi_i(v|_n) = \phi_j(v|_n)$ since $v|_n = m_{n,n}$, and $\phi_i(v|_{c(v)}) > \phi_j(v|_{c(v)})$ since $v|_{c(v)} = v$, $i \succ_v j$, and the Shapley-Shubik index is locally strictly monotonic. This implies that for some k the inequality must be strict.   □

Remark 2. The same argument can be repeated when a player's Shapley-Shubik index is replaced by his number of swings, $d(i, w)$. This, however, does not imply an analogue to Theorem 1 for the Banzhaf index because of the normalization involved in its computation. We suspect that in some cases adding a chamber with a one man - one vote game might increase differences between Banzhaf indices of some players but we do not have a counterexample.

Remark 3. Another interesting open problem is whether Theorem 1 is also true in the more general case when the game v is not complete. Such a result would be attractive because it would imply that adding a chamber with all players equal is a universal tool mitigating power inequalities between players. We believe that this may be true but have no proof; in any case, the method used in our proof of Theorem 1 does not work for incomplete games.

References

1. Banzhaf, J.F.: Weighted voting does not work: a mathematical analysis. Rutgers Law Rev. **19**, 317–343 (1965)
2. Carreras, F., Freixas, J.: Complete simple games. Math. Soc. Sci. **32**, 139–155 (1996)
3. Felsenthal, D., Machover, M., Zwicker, W.: The bicameral postulates and indices of a priori voting power. Theory Decis. **44**, 83–116 (1998)
4. Isbell, J.R.: A class of majority games. Q. J. Math. Oxford Ser. **7**, 183–187 (1956)
5. Shapley, L.S., Shubik, M.: A method for evaluationg the distribution of power in a committee system. Am. Polit. Sci. Rev. **48**, 787–792 (1954)
6. Taylor, A., Zwicker, W.: Simple Games. Princeton University Press, Princeton (1999)

On Ordering a Set of Degressively Proportional Apportionments

Katarzyna Cegiełka, Piotr Dniestrzański, Janusz Łyko, Arkadiusz Maciuk, and Radosław Rudek[✉]

Wrocław University of Economics, Komandorska 118/120, 53-345 Wrocław, Poland
{katarzyna.cegielka,piotr.dniestrzanski,janusz.lyko,
arkadiusz.maciuk,radoslaw.rudek}@ue.wroc.pl

Abstract. The most important problem in a practical implementation of degressive proportionality is its ambiguity. Therefore, we introduce an order relation on a set of degressively proportional allocations. Its main idea is to define greater allocations such that emerge from other after transferring a certain quantity of goods from smaller to greater entities contending in distribution. Thus, maximal elements in this ordering are indicated as the sought-after solution sanctioning boundary conditions as the only reason of moving away from the fundamental principle of proportionality. In case of several maximal elements the choice of one allocation remains an open issue, but the cardinality of the set from which we make a choice can be reduced significantly. In the best-known example of application of degressive proportionality, which is the apportionment of seats in the European Parliament, the considered set contains a maximal element. Thereby, there exits an allocation that is nearest to the proportional distribution with respect to transfer relation.

Keywords: Allocation · Degressive proportionality · Transfer order

1 Introduction

A problem of fair distribution is approached by several disciplines of science in different perspectives. The issue is of interest to philosophers, sociologists, mathematicians, politicians and economists, since there is no unambiguous answer to the question of fairness of a distribution. The answer depends on many factors, with cultural environment in the first place. The concept of fairness in Europe has been formed by Aristotelian principle of proportionality. According to this principle each entity participating in distribution of goods should be allocated a quantity proportional to the value of the entity (participants). The value of entities can be perceived in many ways. In the area of electing political representation it is typically the number of population in an electoral district. When deciding on various issues in joint-stock companies it is capital holdings evidenced by the number of shares owned by shareholders. When determining taxes it is either income or wealth of a given entity.

© Springer International Publishing AG 2017
N.T. Nguyen et al. (Eds.): TCCI XXVII, LNCS 10480, pp. 53–62, 2017.
https://doi.org/10.1007/978-3-319-70647-4_4

A rule of proportional allocation however, is not perceived, even in Europe, as the unique solution to the problem of a fair distribution. In particular, its application can be disapproved when the values of participants differ a lot. In such a case the large value of entities can take unfair advantage and be criticized as socially unjust. This problem can be solved by assuming that the entities with greater value reduce their demands for the benefit of entities with smaller value. Thus a classical proportion is distorted for the sake of increasing the portions allocated to those participants whose proportional shares are smallest, at the cost of those with greater shares.

One possible implementation of the mentioned idea is a degressively proportional rule stating that a greater participants cannot be allocated less than a smaller participants, but the quotient of the amount of a good to the participants value cannot increase with the increase of their value.

Definition 1. *A positive vector of shares* $S = (s_1, s_2, \ldots, s_n)$ *is degressively proportional with respect to positive, nondecreasing sequence of values (demands, claims)* $P = (p_1, p_2, \ldots, p_n)$ *if* $s_1 \leq s_2 \leq \ldots \leq s_n$ *and* $p_1/s_1 \leq p_2/s_2 \leq \ldots \leq p_n/s_n$.

Let us also introduce: $s_1 = m$, $s_n = M$, $\sum_{i=1}^{n} s_i = H$. These three equalities will be henceforth called boundary conditions. In case of degressively proportional allocation they determine the upper and lower bounds for the quantities of goods allocated to participants, and the total quantity of goods.

The most important problem in a practical implementation of degressive proportionality is its ambiguity. Even if we determine the total quantity of an allocated good H, the number of vectors that are degressively proportional with respect to a given sequence of claims can be infinite; an outcome certainly different from the unambiguous solution provided by the rule of proportional allocation. Ambiguity of proportional distribution (allocation) emerges once we distribute indivisible goods, because indivisibility requires s_i to be expressed as natural numbers, thus rounding is necessary, which can be understood in many ways ([7]). In case of divisible goods a sequence of quota $Q = (q_1, q_2, \ldots, q_n)$, $q_i = \frac{p_i H}{\sum_{i=1}^{n} p_j}$, is a unique solution satisfying the rule of proportional distribution.

In this paper, we consider only degressively proportional apportionments in whole numbers with boundary conditions. The problem of ambiguous solutions however is not made simpler. Assuming that $s_1 = m$, $s_n = M$ and $\sum_{i=1}^{n} s_i = H$, we only ensure that the set

$$DP(P, m, M, H) = \Big\{ (s_1, s_2, \ldots, s_n) : s_1 = m \leq s_2 \leq \ldots s_n = M,$$

$$\frac{p_1}{s_1} \leq \frac{p_2}{s_2} \leq \ldots \frac{p_n}{s_n}, \sum_{i=1}^{n} s_i = H \Big\}$$

of all vectors with natural elements, that are degressively proportional with respect to a vector $(p_1, p_2, \ldots, p_n)$, has a finite number of elements. The number of elements in this set can be arbitrarily large; therefore we have to select one out of many possible allocations.

A natural way to obtain uniqueness is by reference to a fundamental principle of distribution, i.e. to proportional allocation. Considering the boundary conditions one can assume that they are the only factors, which lead to a degressively proportional distribution, and that is why one should seek a solution in the set $DP(P, m, M, H)$ that is nearest to the sequence of quota generated by proportional distribution. In addition, after the boundary conditions are determined, an actual solution can be anticipated. In view of the practice where the boundary conditions are often negotiated among participants, we should consider this interpretation as leading to a compromise.

Allocation of seats in the European Parliament is a practical implementation of division consistent with this idea. The rule itself is stated in the Treaty of Lisbon: "*The European Parliament shall be composed of representatives of the Union's citizens. They shall not exceed seven hundred and fifty in number, plus the President. Representation of citizens shall be degressively proportional, with a minimum threshold of six members per Member State. No Member State shall be allocated more than ninety-six seats*" ([9]). The resolution issued by the European Parliament in 2007 specifies among other that "*the minimum and maximum numbers set by the Treaty must be fully utilised to ensure that the allocation of seats in the European Parliament reflects as closely as possible the range of populations of the Member States*" ([5]). A natural interpretation of these words shows that legislators propose to determine an allocation that is nearest to proportional as a result of degressively proportional approach with boundary conditions $m = 6$, $M = 96$, $H = 751$.

Now we need to formally comprehend nearness of a division with respect to a proportional distribution. The literature includes at least two groups of proposals to solve this problem. First, the application of known methods of proportional allocation exemplified by the Cambridge Compromise ([2,3]) and maxprop ([1]). Second, the application of elements of operation research that seek an optimum on the set $DP(P, m, M, H)$ with respect to some criterion (see [6,8]).

In this paper, we propose an approach to finding a solution that differs significantly from the above mentioned. We introduce an order relation on the set $DP(P, m, M, H)$, which is consistent with required proportionality. The main idea of this ordering is to define greater allocations, i.e. such that emerge from other after transferring a certain quantity of goods from smaller to greater entities contending in distribution. Maximal elements in this ordering are indicated as the sought-after solution thus sanctioning boundary conditions as the only reason of moving away from the fundamental principle of proportionality. Maximal allocations in this ordering have the property that entities with greater values minimally reduce their claims subject to constraints imposed by the definition of the set $DP(P, m, M, H)$.

2   Transfer Order

Boundary conditions can be the reason why all elements of the set $DP(P, m, M, H)$ are distant from the proportional distribution. The more distant the values of m and M are from the quota allocated to the smallest ($\frac{p_1 H}{\sum_{i=1}^{n} p_i}$) and greatest ($\frac{p_n H}{\sum_{i=1}^{n} p_i}$) participant, the more remarkable such outcome is.

Example 1. For given $P = (10, 20, 50, 70)$, $H = 15$, $m = 2$, $M = 5$ a proportional distribution is assured by the vector $(1, 2, 5, 7)$. On the other hand, a set of all degressively proportional distributions has two elements: $DP\big((10, 20, 50, 70), 2, 5, 15\big) = \{(2, 4, 4, 5), (2, 3, 5, 5)\}$.

The three greatest participants in Example 1 are allocated at least the quantity of goods in each degressively proportional distribution as under proportional distribution. This happens, of course, at the expense of the subsequent, fourth, greatest contender. The size of this loss and also the gain of the entity with the smallest value are defined by the boundary conditions $m = 2$ and $M = 5$. The two remaining participants can receive either four units of good each or the smaller one – three units, and the greater – five. Therefore a question arises as to which allocation is nearer to the proportional distribution. It is easily seen that in case of the first solution $s_2 = 2q_2$ holds, i.e. the second participants is allocated twice its proportional share, whereas $s_3 < q_3$, which implies that this participant does not benefit if we change the allocation rule from proportional to degressively proportional. We also have $S^* = S + (0, -1, 1, 0)$, where $S = (2, 4, 4, 5)$ and $S^* = (2, 3, 5, 5)$. These relationships can be interpreted accordingly that the allocation S^* means we take one unit of a distributed good from the second and give it to the third participant. Thus it is a transfer of good from a smaller to a greater one. If we wish to come near a proportional allocation as much as possible, the principle of degressively proportional apportionment requires that we transfer as many units of goods as possible from entities with smaller values to entities with greater values. The following two definitions formalize this concept.

Definition 2. *The set of positive transfers is defined as*

$$TR_+ = \Big\{(t_1, t_2, \dots, t_n) : \sum_{i=1}^{n} t_i = 0, \sum_{i=1}^{k} t_i \leq 0 \ for \ k = 1, \dots, n\Big\},$$

whereas its elements are called transfers.

Definition 3. *A positive transfer relation on a nonempty set $DP(P, m, M, H)$ is called a relation $\leq_{TR_+}$ such that*

$$S \leq_{TR_+} S^* \iff (\exists T \in TR_+, S^* = S + T).$$

The relationship $\sum_{i=1}^{n} t_i = 0$ (in Definition 2) ensures that the vectors S and S^* (in Definition 3) have the equal sums of their elements, hence they represent the allocations of the same quantity of goods. The relationship $\sum_{i=1}^{k} t_i \leq 0$ shows the direction of transfers – from smaller to greater entities. It follows from Definition 1 that undervaluation of entities (under degressively proportional distribution) does not decrease along with the increase of their sizes. Thus, if we want to obtain the allocation, which is nearest to the proportional, the transfer of goods in this direction is right.

Henceforth, instead of a positive transfer relation, we shall briefly refer to a transfer relation. With the data from Example 1 we have $S^* = S + T$, where $T = (0, -1, 1, 0) \in TR_+$, hence $S \leq_{TR_+} S^*$ holds.

A transfer relation can be employed to comparisons of degressively proportional distributions with respect to their nearness to the proportional distribution. It is a result of the following proposition.

Proposition 1. *Relation $\leq_{TR_+}$ is a partial order relation.*

Proof. **Reflexivity.** For each $S \in DP(P, m, M, H)$ holds $S = S + \theta_n$, where $\theta_n = (0, 0, \ldots, 0)$, thus we have $S \leq_{TR_+} S$.

Antisymmetry. Let $S \leq_{TR_+} S^*$ and $S^* \leq_{TR_+} S$, then $S^* = S + T_1$ and $S = S^* + T_2$ hold for certain $T_1, T_2 \in TR_+$. Hence we have $S = S + T_1$, or $T_1 = -T_2$. As the elements of vectors T_1 and T_2 are non-positive, then $T_1 = T_2 = \theta_n$ must hold, thereby $S = S^*$.

Transitivity. Let $S \leq_{TR_+} S^*$ and $S^* \leq_{TR_+} S^{**}$, then there exist such $T_1, T_2 \in TR_+$ that $S^* = S + T_1$ and $S^{**} = S^* + T_2$ hold. Therefore, we have $S^{**} = S + (T_1 + T_2)$. Let $T_1 = (t_{1,1}, t_{1,2}, \ldots, t_{1,n})$ and $T_2 = (t_{2,1}, t_{2,2}, \ldots, t_{2,n})$ and we need to show that $T_1 + T_2 \in TR_+$. Since $\sum_{i=1}^n t_{1,i} = 0$, $\sum_{i=1}^k t_{1,i} \leq 0$ and $\sum_{i=1}^n t_{2,i} = 0$, $\sum_{i=1}^k t_{2,i} \leq 0$ hold for $k = 1, \ldots, n$, then we have $\sum_{i=1}^n (t_{1,i} + t_{2,i}) = 0$ and $\sum_{i=1}^k (t_{1,i} + t_{2,i}) \leq 0$, which means that $T_1 + T_2 \in TR_+$. Hence we obtain $S \leq_{TR_+} S^{**}$. $\square$

Relation $\leq_{TR_+}$ orders the set of degressively proportional apportionments. As a consequence of the previous analysis, if $S \leq_{TR_+} S^*$ holds, then we acknowledge that S^* is nearer to the proportional allocation than S. On the other hand, relation $\leq_{TR_+}$ does not linearly order a set of degressively proportional distributions.

Proposition 2. *Transfer relation is not a linear order.*

Proof. It is shown by a counterexample. There is given a set $DP(P, m, M, H)$, where $P = (100, 200, 350, 350, 560, 840, 945)$, $m = 2$, $M = 9$ and $H = 40$. On this basis, we have $DP(P, m, M, H) = \{A, B, C, D\}$, where $A = (2, 3, 5, 5, 8, 8, 9)$, $B = (2, 4, 5, 5, 6, 9, 9)$, $C = (2, 3, 5, 5, 7, 9, 9)$ and $D = (2, 4, 5, 5, 7, 8, 9)$. The corresponding Hasse diagram is given in Fig. 1.

Allocations A and B are incomparable under relation $\leq_{TR_+}$, since $A - B = T = (0, -1, 0, 0, 2, -1, 0)$ and $\sum_{i=1}^5 = 1 > 0$. $\square$

As a consequence of Proposition 2, the degressively proportional allocations from the given partially ordered set (poset) $\left(DP(P, m, M, H), \leq_{TR_+}\right)$ can be incomparable in some cases. Thus, we are not able to determine which relation from incomparable ones is nearer to the proportional allocation. But we are interested precisely in such allocation, which is the nearest one to the proportional allocation (with respect to a transfer order). If it exists, it is the greatest

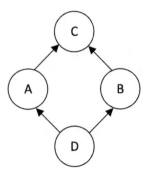

Fig. 1. A Hasse diagram for $A = (2,3,5,5,8,8,9)$, $B = (2,4,5,5,6,9,9)$, $C = (2,3,5,5,7,9,9)$ and $D = (2,4,5,5,7,8,9)$

element of the poset $(DP(P,m,M,H), \leq_{TR_+})$ that will be called a Transfer Order Allocation (TOA). Proposition 3 shows that the greatest element may not exist.

Proposition 3. *There exist P, m, M, H for which a poset $(DP(P,m,M,H), \leq_{TR_+})$ does not contain the greatest element.*

Proof. We prove the proposition by a counterexample. Given the set $DP(P,m,M,H)$, where $P = (100, 200, 466, 466, 931, 1165)$, $m = 2$, $M = 10$, $H = 35$, we have $DP(P,m,M,H) = \{A,B\}$, where $A = (2,3,6,6,8,10)$ and $B = (2,4,5,5,9,10)$. Note that $A - B = T = (0,-1,1,1,-1,0)$, therefore, we have $\sum_{i=1}^{2} t_i = -1$ and $\sum_{i=1}^{4} t_i = 1 > 0$, which means that T is not a transfer (see Definition 2). On the other hand, for $B - A = T' = (0,1,-1,-1,1,0)$, we have $\sum_{i=1}^{2} t_i' = 1$, which is not a transfer. Therefore, allocations A and B are incomparable under the relation $\leq_{TR_+}$, thereby the existence of two maximal elements results in the lack of the greatest element. A related Hasse diagram is given in Fig. 2. □

Fig. 2. An example of a Hasse diagram without the greatest element for $A = (2,3,6,6,8,10)$ and $B = (2,4,5,5,9,10)$

When the greatest element does not exist, the transfer order does not unambiguously point to the allocation from the set $DP(P,m,M,H)$ as the nearest one to the proportional distribution. However, if the greatest element exists, it is easily obtained by Proposition 4. Before expressing this proposition, we shall prove the following lemma.

Lemma 1. *The greatest element in the poset* $(DP(P, m, M, H), \leq_{AL})$, *where* $\leq_{AL}$ *is an antilexicographic order, is the maximal element in the poset* $(DP(P, m, M, H), \leq_{TR_+})$.

Proof. Let us assume the opposite that S is not a maximal element in the set $DP(P, m, M, H)$ with transfer order. Then there exists such element S^* that $S \leq_{TR_+} S^*$ holds. This means that there exists such transfer T that $S^* = S + T$ holds. It follows then from Definition 2 that S^* is also greater than S under antilexicographic order, hence S is not the greatest element in the set $DP(P, m, M, H)$ under antilexicographic order.   □

Proposition 4. *If there exists a greatest element in the set* $DP(P, m, M, H)$ *with transfer order, then it is also the greatest element in the same set with antilexicographic order.*

Proof. It results directly from Lemma 1.   □

3   Case Study

The acts of law, which introduce and regulate the distribution of seats in the European Parliament, determine the set $DP(P, 6, 96, 751)$, where $P = (p_1, p_2, \ldots, p_n)$ is the vector of populations of the Member States. There are 751 seats, which are distributed among 28 states in a degressively manner, with the least populated country receiving 6 seats and the most populated – 96 seats. Obviously, as a result of the assumed boundary conditions countries with less population are allocated more seats than under proportional allocation, and countries with more population are given fewer seats than under proportional scheme. For example, proportional allocation would assign no more than one seat to small countries such as Malta, Luxembourg or Cyprus, while Germany would be assigned at least 120 seats. However the smallest country and the greatest country are allocated precisely 6 and 96 seats respectively, the numbers of seats for remaining states are not determined uniquely. What is more, with data on populations in 2012 (employed to determine the composition of the European Parliament in the current term of 2014–2019), the set $DP(P, 6, 96, 751)$ has cardinality more than 5 million (see [6]). Undoubtedly therefore, a rule must be indicated that will allow the choice of a specific allocation.

It turns out that the set $DP(P, 6, 96, 751)$ contains the greatest element with respect to transfer order, i.e. the transfer order proposed in the previous section allows finding the allocation that is nearest to the proportional one. This allocation denoted by TOA and presented in column 9 of Table 1 reduces the undervaluation, imposed by the definition of the set $DP(P, m, M, H)$, of most populated countries for the benefit of less populated countries compared to proportional allocation. Thus in contrast to other presented allocations, TOA gives more populated countries more seats. Therefore it ensures both the satisfaction of postulates of parliamentarians requiring the best possible representation of differences in populations of the Member States as well as the compliance with current regulations.

Table 1. Examples of the allocations of seats in the European Parliament

#	Country	Population [in thousands]	Current		Cambridge Compromise		LaRSA	TOA	
i		p_i	s_i	p_i/s_i	s_i	p_i/s_i	s_i	s_i	p_i/s_i
1	2	3	4	5	6	7	8	9	10
1	Malta	416.11	6	2.48	6	2.48	6	6	2.48
2	Luxembourg	524.85	6	3.12	6	3.12	6	6	3.12
3	Cyprus	862.01	6	5.13	7	4.40	6	6	5.13
4	Estonia	1339.66	6	7.97	7	6.84	6	6	7.97
5	Latvia	2041.76	8	9.12	8	9.12	6	6	12.15
6	Slovenia	2055.50	8	9.18	8	9.18	6	6	12.24
7	Lithuania	3007.76	11	9.77	9	11.94	7	7	15.35
8	Croatia	4398.15	11	14.28	11	14.28	10	10	15.71
9	Ireland	4582.77	11	**14.88**	11	14.88	10	10	16.37
10	Finland	5401.27	13	14.84	12	16.08	11	11	17.54
11	Slovakia	5404.32	13	14.85	12	16.08	11	11	17.55
12	Denmark	5580.52	13	15.33	12	16.61	11	11	18.12
13	Bulgaria	7327.22	17	15.39	14	18.69	14	14	18.69
14	Austria	8443.02	19	15.87	16	18.85	16	16	18.85
15	Sweden	9482.86	19	**17.82**	17	19.92	17	17	19.92
16	Hungary	9957.73	21	16.93	17	**20.92**	17	17	20.92
17	Czech Rep	10505.45	21	17.87	18	**20.84**	17	17	22.07
18	Portugal	10541.84	21	17.93	18	**20.92**	17	17	22.15
19	Belgium	11041.27	21	18.78	19	20.75	17	17	23.20
20	Greece	11290.94	21	19.20	19	21.22	17	17	23.72
21	Netherlands	16730.35	26	22.98	25	23.90	24	24	24.90
22	Romania	21355.85	32	23.83	31	24.60	30	30	25.42
23	Poland	38538.45	51	26.99	51	26.99	54	54	25.49
24	Spain	46196.28	54	**30.55**	60	27.50	64	64	25.78
25	Italy	60820.76	73	29.76	78	27.85	83	83	26.17
26	UK	62989.55	73	**30.82**	80	28.12	85	85	26.47
27	France	65397.91	74	**31.56**	83	28.14	87	87	26.85
28	Germany	81843.74	96	30.45	96	30.45	96	96	30.45
	Total	508077.9	751		751		751	751	

Table 1 presents several selected allocations. Columns 4 and 5 contain numbers of seats allocated to countries for the current term of the European Parliament in 2014–2019. Bold entries in column 5 indicate violations of the degressive proportionality principle. Nevertheless, the parliamentarians agreed to the incompliance of this solution with currently binding law (see [4]).

Column 6 shows the distribution in compliance with the Cambridge Compromise ([2]), which is determined by the rule $s_i = \max\{5 + \lceil p_i/a \rceil\}$, where a constant a is set to ensure that the total of all distributed seats equals 751 (for example $a \in [839.94, 844.30]$ for data from 2012). This distribution, similarly as current allocation, violates the principle of degressive proportionality (bold entries in column 7). This incompliance results from the design of the procedure that can return an allocation, which is not an element of the set $DP(P, 6, 96, 751)$ (for more details see [2]). The allocations presented in columns 8 and 9 are free of that flaw.

Column 8 contains the allocation generated by LaRSA algorithm ([6]), which consists in searching the set $DP(P, 6, 96, 751)$ and selecting the allocation that minimizes the sum of squared distances form the proportional allocation, i.e. $\sum_{i=1}^{n} \left(p_i H / \sum_{j=1}^{n} p_j \right) \rightarrow \min$ .Column 9 contains the TOA, the greatest element of the set $DP(P, 6, 96, 751)$ with transfer order. By Proposition 4, this allocation is also the greatest element in the set $DP(P, 6, 96, 751)$ with antilexicographic order. As we can see, the LaRSA and TOA allocations are identical. However that is not necessarily the case. There are some examples showing that the allocations generated by those methods are different. In case of the European Parliament though, the equality of those two allocations additionally supports the idea of coming closer to the proportional allocation by means of transfer of seats. Obviously both approaches ensure that the degressively proportional allocation is obtained, because they are selected from the set $DP(P, 6, 96, 751)$.

4   Conclusions

A transfer relation orders a given set of degressively proportional allocations. Except for cases when it is not a linear order, the relation makes it possible to specify which allocation is nearer to the proportional distribution, and particularly to determine the maximal elements, i.e. such that no other distribution is better. If there is a greatest element in the given set $DP(P, m, M, H)$ with transfer order, we assume that this is the optimal allocation. In addition, there is an alternative method to find it – by Proposition 4 it is the greatest element in the set $DP(P, m, M, H)$ with antilexicographic order. In case of several maximal elements the choice of one distribution remains an open issue, but the cardinality of the set from which we make a choice can be reduced significantly. In the best-known example of application of degressive proportionality, i.e. the apportionment of seats in the European Parliament, the set $DP(P, m, M, H)$ contains a maximal element. In other words, there exits an allocation that is nearest to the proportional distribution with respect to transfer relation.

Acknowledgement. The results presented in this paper have been supported by the Polish National Science Centre under grant no. DEC-2013/09/B/HS4/02702.

References

1. Dniestrzański, P., Łyko, J.: Proportionality in the issue of distribution of seats in the European Parliament. In: Proceedings of the 4th International Conference on Computer Science & Computational Mathematics, pp. 542–546 (2015)
2. Grimmett, G., Laslier, J.-F., Pukelsheim, F., Ramirez-González, V., Rose, R., Słomczyński, W., Zachariasen, M., Życzkowski, K.: The allocation between the EU member states of the seats in the European Parliament. European Parliament, Directorate-General for Internal Policies, Policy Department C: Citizen's Rights and Constitutional Affairs, PE 432.760 (2011)
3. Grimmett, G.R.: European apportionment via the Cambridge Compromise. Math. Soc. Sci. **63**, 68–73 (2012)
4. Gualtieri, R., Trzaskowski, R.: Report on the composition of the European Parliament with a view to the 2014 elections. A7–0041/2013 (2013)
5. Lamassoure, A., Severin, A.: Report on the composition of the European Parliament. A6–0351/2007 (2007)
6. Łyko, J., Rudek, R.: A fast exact algorithm for the allocation of seats for the EU Parliament. Expert Syst. Appl. **40**, 5284–5291 (2013)
7. Pukelsheim, F.: Proportional Representation. Apportionment Methods and Their Applications. Springer, Cham (2014)
8. Serafini, P.: Allocation of the EU Parliament seats via integer linear programming and revised quotas. Math. Soc. Sci. **63**, 107–113 (2012)
9. Treaty of Lisbon. Official Journal of the European Union, C 306, **50** (2007)

Preorders in Simple Games

Josep Freixas and Montserrat Pons[(✉)]

Department of Mathematics, Universitat Politècnica de Catalunya, Manresa, Spain
{josep.freixas,montserrat.pons}@upc.edu

Abstract. Any power index defines a total preorder in a simple game and, thus, induces a hierarchy among its players. The desirability relation, which is also a preorder, induces the same hierarchy as the Banzhaf and the Shapley indices on linear games, i.e., games in which the desirability relation is total. The desirability relation is a sub–preorder of another preorder, the weak desirability relation, and the class of weakly linear games, i.e., games for which the weak desirability relation is total, is larger than the class of linear games. The weak desirability relation induces the same hierarchy as the Banzhaf and the Shapley indices on weakly linear games. In this paper, we define a chain of preorders between the desirability and the weak desirability preorders. From them we obtain new classes of totally preordered games between linear and weakly linear games.

Keywords: Simple game · Power index · Preorder · Desirability · Weak desirability · Linear game · Weakly linear game

1 Introduction

Any power index considered in a simple game induces a total preorder on the set of players, and, thus, a hierarchy among them. Two power indices which induce the same hierarchy in a simple game are said to be *ordinally equivalent* in it. In this paper we refer to three power indices: the Shapley–Shubik index (SS, henceforth) [14,15], the Penrose–Banzhaf–Coleman index (PBC, henceforth) [1, 4,12] and the Johnston index [11]. It is known [5,13] that the PBC, the SS and the Johnston power indices are ordinally equivalent in linear games, and that the common induced hierarchy is the one given by the desirability relation.

In [3] weakly linear games were introduced and it was proved that all regular semivalues (i.e., semivalues with positive coefficients for the marginal contributions, see [2,3]) are ordinally equivalent for this kind of games, and that the common induced hierarchy is the one given by the weak desirability relation. As every linear simple games is weakly linear, and both the PBC and the SS power indices are regular semivalues, this work extends and generalizes the former ones in relation with these two indices. The ordinal equivalence of the SS, the PBC and the Johnston indices in a class larger than linear games but smaller than weakly linear games was proved in [7].

© Springer International Publishing AG 2017
N.T. Nguyen et al. (Eds.): TCCI XXVII, LNCS 10480, pp. 63–73, 2017.
https://doi.org/10.1007/978-3-319-70647-4_5

In this paper, a chain of families of simple games, between linear and weakly linear games, are defined and some of their properties are studied. In all these classes of simple games, the Banzhaf and the Shapley–Shubik indices are ordinally equivalent. The smallest family in this chain is the class of linear games and the largest one is the class of weakly linear games. A somehow similar work was developed in [16] by defining a chain of classes of simple games between weighted simple games and linear simple games.

The paper is organized as follows. Basic definitions and preliminary results are included in Sect. 2. Section 3 contains new characterizations of the desirability and the weak desirability relations. In Sect. 4 the m–desirability relations are defined and some of their properties are studied. Some Conclusions end the paper in Sect. 5.

2   Definitions and Preliminaries

In the sequel, $N = \{1, 2, \ldots, n\}$ denote a fixed but otherwise arbitrary finite set of *players*. Any subset $S \subseteq N$ is a *coalition*. A *simple game* v (in N, omitted hereafter) is a cooperative game, i.e., a function $v : 2^N \to \mathbb{R}$ with $v(\emptyset) = 0$, such that: (a) $v(S) = 0$ or 1 for any coalition S,[1] (b) v is monotonic, i.e., $v(S) \leq v(T)$ whenever $S \subset T$, and (c) $v(N) = 1$. Either the family of *winning* coalitions $\mathcal{W} = \mathcal{W}(v) = \{S \subseteq N : v(S) = 1\}$ or the subfamily of *minimal* winning coalitions $\mathcal{W}^m = \mathcal{W}^m(v) = \{S \in \mathcal{W} : T \subset S \Rightarrow T \notin \mathcal{W}\}$ determines a simple game.

Given a simple game v, let us consider, for each $i \in N$, and for every integer h with $1 \leq h \leq n$, some important subsets of N:

$$\mathcal{P}_i = \{S \subseteq N : i \in S\} \quad \text{and} \quad \mathcal{P}_i(h) = \{S \in \mathcal{P}_i : |S| = h\}.$$

$\mathcal{P}_i$ is the set of coalitions S that contain i, while $\mathcal{P}_i(h)$ is the subset of such coalitions having cardinality h.

$$\mathcal{W}_i = \{S \in \mathcal{W} : i \in S\} \quad \text{and} \quad \mathcal{W}_i(h) = \{S \in \mathcal{W}_i : |S| = h\}.$$

$\mathcal{W}_i$ is the set of winning coalitions S that contain i, while $\mathcal{W}_i(h)$ is the subset of such coalitions having cardinality h.

$$\mathcal{C}_i = \{S \in \mathcal{W}_i : S \setminus \{i\} \notin \mathcal{W}\} \quad \text{and} \quad \mathcal{C}_i(h) = \{S \in \mathcal{C}_i : |S| = h\}.$$

$\mathcal{C}_i$ is the set of winning coalitions S that are crucial for i, while $\mathcal{C}_i(h)$ is the subset of such coalitions having cardinality h. It is obvious that

$$\mathcal{C}_i(h) \subseteq \mathcal{W}_i(h) \subseteq \mathcal{P}_i(h).$$

Notice that for $h = 1$ the set $\mathcal{P}_i(1)$ only contains the singleton $\{i\}$, $\mathcal{W}_i(1) = \mathcal{C}_i(1)$ and $\{i\} \in \mathcal{C}_i(1) \Leftrightarrow \{i\} \in \mathcal{W}$. On the other extreme, for $h = n$ the set $\mathcal{P}_i(n)$ only contains the total set N, $\mathcal{W}_i(n) = \mathcal{P}_i(n)$ and $N \in \mathcal{C}_i(n) \Leftrightarrow N \setminus \{i\} \notin \mathcal{W}$.

---

[1] For a detailed discussion of some issues raised by allowing abstentions, see Felsenthal and Machover [6] and for several levels of approval in input and output, see Freixas and Zwicker [9].

Definition 1. *The desirability relation* ([10])
Let v be a simple game and $i, j \in N$. Then

$$\begin{aligned}
i \succsim_D j &\quad\Leftrightarrow\quad \{\, S \cup \{j\} \in \mathcal{W} \Rightarrow S \cup \{i\} \in \mathcal{W} \,\} \text{ for any } S \subseteq N \backslash \{i, j\}, \\
i \succ_D j &\quad\Leftrightarrow\quad i \succsim_D j \quad \text{and} \quad j \not\succsim_D i, \\
i \approx_D j &\quad\Leftrightarrow\quad i \succsim_D j \quad \text{and} \quad j \succsim_D i.
\end{aligned}$$

It is well known that $\succsim_D$ is a preordering. The relation $\succsim_D$ (resp., $\succ_D$) is called the desirability (resp., strict desirability) relation, and $\approx_D$ is the equi–desirability relation.

Definition 2. *Linear game*[2]
A simple game v is linear *whenever the desirability relation $\succsim_D$ is complete.*

Definition 3. *The weak desirability relation* ([3])
Let v be a simple game and $i, j \in N$. Then

$$\begin{aligned}
i \succsim_d j &\quad\Leftrightarrow\quad |\mathcal{C}_i(h)| \geq |\mathcal{C}_j(h)| \quad \text{for any } h \text{ with } 1 \leq h \leq n, \\
i \succ_d j &\quad\Leftrightarrow\quad i \succsim_d j \quad \text{and} \quad j \not\succsim_d i, \\
i \approx_d j &\quad\Leftrightarrow\quad i \succsim_d j \quad \text{and} \quad j \succsim_d i.
\end{aligned}$$

Then $\succsim_d$ is a preordering called the weak desirability relation. The relation $\succ_d$ is the strict weak desirability relation and $\approx_d$ is the weak equi–desirability relation.

In [5] it is proved that the desirability relation is a sub–preordering of the weak desirability relation, that is to say, for any $i, j \in N$, $i \succsim_D j$ implies $i \succsim_d j$ and $i \succ_D j$ implies $i \succ_d j$.

Definition 4. *Weakly linear game* ([3])
A simple game v is weakly linear *whenever the weak desirability relation $\succsim_d$ is complete.*

As stated in [3], the completeness of the desirability relation $\succsim_D$ implies the completeness of the weak desirability relation $\succsim_d$ so that all linear games are also weakly linear.

Moreover, if v is a linear simple game then v is weakly linear and the desirability relation $\succsim_D$ and the weak desirability relation $\succsim_d$ coincide.

3   Other Characterizations of the Desirability and the Weak Desirability Relations

Given two different elements $i, j \in N$ we can establish a function

$$\varphi_{ji} : \mathcal{P}_j \to \mathcal{P}_i$$

---

[2] Linear games are also called complete, ordered or directed games in the literature, see Taylor and Zwicker [16] for references on these names.

defined by:

$$\varphi_{ji}(S) = \begin{cases} S & \text{if } i \in S \\ (S \setminus \{j\}) \cup \{i\} & \text{if } i \notin S \end{cases}$$

It is not difficult to see that φ_{ji} is bijective and its inverse is φ_{ij}. Notice that, for any $S \in \mathcal{P}_j$ its image $\varphi_{ji}(S)$ always contain $S \setminus \{j\}$.

In the following subsections we study the restrictions of φ_{ji} to $\mathcal{W}_j(h)$ and to $\mathcal{C}_j(h)$. We will see that we can characterize the relations $i \succsim_D j$ and $i \succsim_d j$ by using these restrictions.

3.1   The Restriction of φ_{ji} to $\mathcal{W}_j(h)$

It is clear that, for any $S \in \mathcal{W}_j(h)$, $\varphi_{ji}(S) \in \mathcal{P}_i(h)$, but it is not always true (except for $h = n$) that $\varphi_{ji}(S) \in \mathcal{W}_i(h)$. The following proposition gives a characterization of this fact.

Proposition 1. *Let v be a simple game, i and j be different elements in N and h be an integer with $1 \le h < n$. Then,*

$$\varphi_{ji}(\mathcal{W}_j(h)) \subseteq \mathcal{W}_i(h) \Leftrightarrow \begin{cases} S \cup \{j\} \in W \Rightarrow S \cup \{i\} \in W \\ \\ \text{for any } S \subseteq N \setminus \{i,j\} \text{ with } |S| = h - 1. \end{cases}$$

Proof:
Assume that $\varphi_{ji}(\mathcal{W}_j(h)) \subseteq \mathcal{W}_i(h)$, and let $S \subseteq N \setminus \{i,j\}$ be such that $S \cup \{j\} \in W$ and $|S| = h - 1$. Taking $T = S \cup \{j\}$ it is $T \in \mathcal{W}_j(h)$ and, thus, $\varphi_{ji}(T) \in W$ by hypothesis. But, since $i \notin T$, it is $\varphi_{ji}(T) = (T \setminus \{j\}) \cup \{i\} = S \cup \{i\}$. This proves that $S \cup \{i\} \in W$.
Conversely, assume the hypothesis and let $S \in \mathcal{W}_j(h)$. To prove that $\varphi_{ji}(S) \in \mathcal{W}_i(h)$ we only need to see that $\varphi_{ji}(S) \in W$. There are two possibilities: a) If $i \in S$ then $\varphi_{ji}(S) = S$ and, obviously, $\varphi_{ji}(S) \in W$. b) If $i \notin S$ then $\varphi_{ji}(S) = (S \setminus \{j\}) \cup \{i\}$. Now, taking $T = S \setminus \{j\}$ it is $T \subseteq N \setminus \{i,j\}$, $|T| = h - 1$ and $T \cup \{j\} = S \in W$, thus, $T \cup \{i\} = \varphi_{ji}(S) \in W$.   □

Proposition 2. *Let v be a simple game, i and j be different elements in N. Then,*

$$i \succsim_D j \Leftrightarrow \varphi_{ji}(\mathcal{W}_j(h)) \subseteq \mathcal{W}_i(h) \quad \text{for any integer } h \text{ with } 1 \le h \le n$$
$$i \approx_D j \Leftrightarrow \varphi_{ji}(\mathcal{W}_j(h)) = \mathcal{W}_i(h) \quad \text{for any integer } h \text{ with } 1 \le h \le n.$$

Proof:
From Definition 1 it is $i \succsim_D j$ if and only if $S \cup \{j\} \in W \Rightarrow S \cup \{i\} \in W$ for any $S \subseteq N \setminus \{i,j\}$. Since $|S| \le n - 2$, from Proposition 1 this last assertion is satisfied if and only if $\varphi_{ji}(\mathcal{W}_j(h)) \subseteq \mathcal{W}_i(h)$ for any h with $1 \le h \le n - 1$. But it is always true that $\varphi_{ji}(\mathcal{W}_j(n)) \subseteq \mathcal{W}_i(n)$ and this proves the first part. The second part is an obvious consequence of the first one.   □

3.2 The Restriction of φ_{ji} to $\mathcal{C}_j(h)$

Proposition 3. *Let v be a simple game, i and j be different elements in N and h be an integer with $1 < h \leq n$. Then,*

$$\varphi_{ji}(\mathcal{C}_j(h)) \subseteq \mathcal{C}_i(h) \quad \Leftrightarrow \quad \begin{cases} \varphi_{ji}(\mathcal{W}_j(h)) \subseteq \mathcal{W}_i(h) \\ \\ \varphi_{ji}(\mathcal{W}_j(h-1)) \subseteq \mathcal{W}_i(h-1). \end{cases}$$

Proof:

(i) Assume that $\varphi_{ji}(\mathcal{C}_j(h)) \subseteq \mathcal{C}_i(h)$. To prove that $\varphi_{ji}(\mathcal{W}_j(h)) \subseteq \mathcal{W}_i(h)$ consider $S \in \mathcal{W}_j(h)$. If $S \setminus \{j\} \notin W$ then $S \in \mathcal{C}_j(h)$ and, by the hypothesis, $\varphi_{ji}(S) \in \mathcal{C}_i(h) \subseteq \mathcal{W}_i(h)$. Otherwise, if $S \setminus \{j\} \in W$ then $\varphi_{ji}(S) \in W$ because of the monotonicity of v taking into account that $\varphi_{ji}(S) \supseteq S \setminus \{j\}$. Thus, in either case, $\varphi_{ji}(S) \in \mathcal{W}_i(h)$.

To prove that $\varphi_{ji}(\mathcal{W}_j(h-1)) \subseteq \mathcal{W}_i(h-1)$ consider $S \in \mathcal{W}_j(h-1)$. If $i \in S$ then, obviously, $\varphi_{ji}(S) = S \in W$. Otherwise, if $i \notin S$ then $\varphi_{ji}(S) = (S \setminus \{j\}) \cup \{i\}$. If $\varphi_{ji}(S) \notin W$ it would be $S \cup \{i\} \in \mathcal{C}_j(h)$ because $|S \cup \{i\}| = h - 1$, $S \cup \{i\} \in W$, by monotonicity, and $(S \cup \{i\}) \setminus \{j\} = \varphi_{ji}(S) \notin W$. Thus, by hypothesis, $\varphi_{ji}(S \cup \{i\}) \in \mathcal{C}_i(h)$. But, since $\varphi_{ji}(S \cup \{i\}) = (S \cup \{i\})$, by monotonicity it would be $S \notin W$ which is a contradiction. Thus, in either case, $\varphi_{ji}(S) \in \mathcal{W}_i(h-1)$.

(ii) Conversely, assume that $\varphi_{ji}(\mathcal{W}_j(h)) \subseteq \mathcal{W}_i(h)$ and $\varphi_{ji}(\mathcal{W}_j(h-1)) \subseteq \mathcal{W}_i(h-1)$, and consider $S \in \mathcal{C}_j(h)$. To prove that $\varphi_{ji}(S) \in \mathcal{C}_i(h)$ we only need to prove that $\varphi_{ji}(S) \setminus \{i\} \notin W$, because by the hypothesis $\varphi_{ji}(S) \in \mathcal{W}_i(h)$. If $i \notin S$ then $\varphi_{ji}(S) = (S \setminus \{j\}) \cup \{i\}$ and, thus, $\varphi_{ji}(S) \setminus \{i\} = S \setminus \{j\} \notin W$ by the hypothesis. Otherwise, if $i \in S$ then $\varphi_{ji}(S) = S$. In this case, if $\varphi_{ji}(S) \setminus \{i\} \in W$ it would be $\varphi_{ji}(S) \setminus \{i\} \in \mathcal{W}_j(h-1)$ because $\varphi_{ji}(S) \setminus \{i,j\} \notin W$ (by monotonicity, since $(\varphi_{ji}(S) \setminus \{j\} \notin W)$. Thus, by the hypothesis, $\varphi_{ji}(S)(\varphi_{ji}(S) \setminus \{i\}) \in \mathcal{W}_i(h-1)$. But $\varphi_{ji}(S)(\varphi_{ji}(S) \setminus \{i\}) = \varphi_{ji}(S) \setminus \{i\} \notin W$, which is a contradiction. $\square$

Proposition 4. *Let v be a simple game, i and j be different elements in N and h be an integer with $1 < h < n$. Then,*

$$\left. \begin{array}{l} S \cup \{j\} \in W \Rightarrow S \cup \{i\} \in W \\ \\ \text{for any } S \subseteq N \setminus \{i,j\} \text{ with } h - 2 \leq |S| \leq h - 1 \end{array} \right\} \Leftrightarrow \varphi_{ji}(\mathcal{C}_j(h)) \subseteq \mathcal{C}_i(h).$$

Proof:

The property "$S \cup \{j\} \in W \Rightarrow S \cup \{i\} \in W$ for any $S \subseteq N \setminus \{i,j\}$ with $h - 2 \leq |S| \leq h - 1$" is equivalent to "$\varphi_{ji}(\mathcal{W}_j(h)) \subseteq \mathcal{W}_i(h)$ and $\varphi_{ji}(\mathcal{W}_j(h-1)) \subseteq \mathcal{W}_i(h-1)$" from Proposition 1, and, from Proposition 3, this is equivalent to $\varphi_{ji}(\mathcal{C}_j(h)) \subseteq \mathcal{C}_i(h)$. $\square$

Corollary 1. *Let v be a simple game, i and j be different elements in N and h be an integer with $1 < h < n$. Then,*

$$\left. \begin{array}{l} S \cup \{j\} \in W \Rightarrow S \cup \{i\} \in W \\ \\ \text{for any } S \subseteq N \setminus \{i,j\} \text{ with } h - 2 \leq |S| \leq h - 1 \end{array} \right\} \Rightarrow |\mathcal{C}_j(h)| \leq |\mathcal{C}_i(h)|.$$

Corollary 2. *Let v be a simple game, i and j be different elements in N and h be an integer with $1 < h < n$. Then,*

$$\left.\begin{array}{l} \varphi_{ji}(\mathcal{C}_j(h-1)) \subseteq \mathcal{C}_i(h-1) \\[2mm] \varphi_{ji}(\mathcal{C}_j(h+1)) \subseteq \mathcal{C}_i(h+1) \end{array}\right\} \Rightarrow \varphi_{ji}(\mathcal{C}_j(h)) \subseteq \mathcal{C}_i(h).$$

Proof:
From Proposition 4, $\varphi_{ji}(\mathcal{C}_j(h-1)) \subseteq \mathcal{C}_i(h-1)$ implies that $S \cup \{j\} \in \mathcal{W} \Rightarrow S \cup \{i\} \in \mathcal{W}$ for any $S \subseteq N\backslash\{i,j\}$ with $|S| = h-2$. Similarly, $\varphi_{ji}(\mathcal{C}_j(h+1)) \subseteq \mathcal{C}_i(h+1)$ implies that $S \cup \{j\} \in \mathcal{W} \Rightarrow S \cup \{i\} \in \mathcal{W}$ for any $S \subseteq N\backslash\{i,j\}$ with $|S| = h-1$. And the two conditions together are equivalent to $\varphi_{ji}(\mathcal{C}_j(h)) \subseteq \mathcal{C}_i(h)$, again from Proposition 4. $\qquad\qquad\square$

Proposition 5. *Let v be a simple game, i and j be different elements in N. Then,*

$$i \succsim_D j \Leftrightarrow \varphi_{ji}(\mathcal{C}_j(h)) \subseteq \mathcal{C}_i(h) \ \text{ for any integer } h \text{ with } 1 \leq h \leq n$$

$$i \approx_D j \Leftrightarrow \varphi_{ji}(\mathcal{C}_j(h)) = \mathcal{C}_i(h) \ \text{ for any integer } h \text{ with } 1 \leq h \leq n.$$

Proof:
From Proposition 2 it is $i \succsim_D j$ if and only if $\varphi_{ji}(\mathcal{W}_j(h)) \subseteq \mathcal{W}_i(h)$ for any h with $1 \leq h \leq n$. From Proposition 3, this is satisfied if and only if $\varphi_{ji}(\mathcal{C}_j(h)) \subseteq \mathcal{C}_i(h)$ for any h with $1 < h \leq n$. And, taking into account that $\mathcal{W}_k(1) = \mathcal{C}_k(1)$ for any $k \in N$, the first part is proved. The second part is an obvious consequence of the first one. $\qquad\qquad\square$

Proposition 6. *Let v be a simple game, i and j be different elements in N. Then the following assertions are equivalent:*

(i) $\varphi_{ji}(\mathcal{C}_j(n)) \subseteq \mathcal{C}_i(n)$
(ii) $|\mathcal{C}_j(n)| \leq |\mathcal{C}_i(n)|$
(iii) $\varphi_{ji}(\mathcal{W}_j(n-1)) \subseteq \mathcal{W}_i(n-1)$
(iv) $N \setminus \{i\} \in \mathcal{W} \Rightarrow N \setminus \{j\} \in \mathcal{W}$

Proposition 7. *Let v be a simple game, i and j be different elements in N. Then the following assertions are equivalent:*

(i) $\varphi_{ji}(\mathcal{C}_j(1)) \subseteq \mathcal{C}_i(1)$
(ii) $|\mathcal{C}_j(1))| \leq |\mathcal{C}_i(1)|$
(iii) $\{j\} \in \mathcal{W} \Rightarrow \{i\} \in \mathcal{W}$

4   The m–desirability relations

We are going to introduce a collection of new preorders in N.

Definition 5. *Let v be a simple game and $i, j \in N$.*
For any integer m with $1 \leq m \leq n$ we define:

$$i \gtrsim_m j \quad \Leftrightarrow \quad \begin{cases} |\mathcal{C}_j(h)| \leq |\mathcal{C}_i(h)| \quad \text{for any } h \text{ with } m \leq h \leq n, \text{ and} \\ \varphi_{ji}(\mathcal{C}_j(h)) \subseteq \mathcal{C}_i(h) \text{ for any } h \text{ with } 1 \leq h < m. \end{cases}$$

$$i \succ_m j \quad \Leftrightarrow \quad i \gtrsim_m j \quad \text{and} \quad j \not\gtrsim_m i,$$

$$i \approx_m j \quad \Leftrightarrow \quad i \gtrsim_m j \quad \text{and} \quad j \gtrsim_m i.$$

Then $\gtrsim_m$ is called the m–desirability relation. The relation $\succ_m$ is the strict m–desirability relation and $\approx_m$ is the m–equi–desirability relation.

Notice that, using Proposition 4, the second part of the definition of the m–desirability relation can be reformulated in the following way:

Remark 1. For any m with $1 < m \leq n$ it is

$$\left\{ \begin{array}{l} \varphi_{ji}(\mathcal{C}_j(h)) \subseteq \mathcal{C}_i(h) \\[1mm] \text{for any } h \text{ with } 1 \leq h < m \end{array} \right\} \Leftrightarrow \left\{ \begin{array}{l} S \cup \{j\} \in \mathcal{W} \Rightarrow S \cup \{i\} \in \mathcal{W} \\[1mm] \text{for any } S \subseteq N \backslash \{i,j\} \text{ with } |S| \leq m - 2 \end{array} \right\}$$

The following proposition states that the desirability relation and the weak desirability relation are particular cases of m-desirability relations, and that different values of m can give the same m–desirability relation.

Proposition 8. *Let v be a simple game and $i, j \in N$. Then,*

$$i \gtrsim_d j \quad \Leftrightarrow \quad i \gtrsim_2 j \quad \Leftrightarrow \quad i \gtrsim_1 j.$$

$$i \gtrsim_D j \quad \Leftrightarrow \quad i \gtrsim_n j \quad \Leftrightarrow \quad i \gtrsim_{n-1} j.$$

Proof:

From Definition 5, it is clear that $i \gtrsim_1 j$ if and only if $|\mathcal{C}_j(h)| \leq |\mathcal{C}_i(h)|$ for any h with $1 \leq h \leq n$ and, from Definition 5, this is equivalent to $i \gtrsim_d j$. On the other hand, $i \gtrsim_2 j$ if and only if $|\mathcal{C}_j(h)| \leq |\mathcal{C}_i(h)|$ for any h with $2 \leq h \leq n$ and $\varphi_{ji}(\mathcal{C}_j(1)) \subseteq \mathcal{C}_i(1)$. But this last condition is equivalent to $|\mathcal{C}_j(1)| \leq |\mathcal{C}_i(1)|$ and, thus, $i \gtrsim_2 j \Leftrightarrow i \gtrsim_d j$.

Similarly, from Definition 5, $i \gtrsim_n j$ if and only if $|\mathcal{C}_j(n)| \leq |\mathcal{C}_i(n)|$ and $\varphi_{ji}(\mathcal{C}_j(h)) \subseteq \mathcal{C}_i(h)$ for any h with $1 \leq h < n$. But it is clear that $|\mathcal{C}_j(n)| \leq |\mathcal{C}_i(n)|$ if and only if $\varphi_{ji}(\mathcal{C}_j(n)) \subseteq \mathcal{C}_i(n)$ and, using Proposition 5, $i \gtrsim_n j \Leftrightarrow i \gtrsim_D j$. Finally, $i \gtrsim_{n-1} j$ if and only if $|\mathcal{C}_j(n)| \leq |\mathcal{C}_i(n)|$, $|\mathcal{C}_j(n-1)| \leq |\mathcal{C}_i(n-1)|$ and $\varphi_{ji}(\mathcal{C}_j(h)) \subseteq \mathcal{C}_i(h)$ for any h with $1 \leq h < n-1$. From Proposition 5 it is obvious that $i \gtrsim_D j \Rightarrow i \gtrsim_{n-1} j$. Conversely, if $i \gtrsim_{n-1} j$, since $|\mathcal{C}_j(n)| \leq |\mathcal{C}_i(n)|$ is equivalent to $\varphi_{ji}(\mathcal{C}_j(n)) \subseteq \mathcal{C}_i(n)$, applying Corollary 2 for $h = n - 1$ we get $\varphi_{ji}(\mathcal{C}_j(n-1)) \subseteq \mathcal{C}_i(n-1)$ and, thus, $i \gtrsim_{n-1} j \Rightarrow i \gtrsim_D j$. $\qquad\square$

Proposition 9. *Let v be a simple game. Then, for any m with $1 \le m \le n$, the m–desirability relation is a preorder in N.*

Proof:
It is already known that the desirability relation and the weak desirability relation are preorders, so that we can assume $2 < m < n - 1$. To prove that the m–desirability relation is transitive, let i, j, k be different elements in N such that $i \succsim_m j \succsim_m k$. It is clear that $|\mathcal{C}_i(h)| \ge |\mathcal{C}_j(h)| \ge |\mathcal{C}_k(h)|$ for any h with $m \le h \le n$, and thus $|\mathcal{C}_i(h)| \ge |\mathcal{C}_k(h)|$ for any h with $m \le h \le n$. To prove that $\varphi_{ki}(\mathcal{C}_k(h)) \subseteq \mathcal{C}_i(h)$ for any h with $1 \le h < m$ we will see, using Remark 1, that $S \cup \{k\} \in \mathcal{W} \Rightarrow S \cup \{i\} \in \mathcal{W}$, for any $S \subseteq N \backslash \{i, k\}$ with $|S| \le m - 2$. Thus, suppose that $S \subseteq N \backslash \{i, k\}$ is such that $|S| \le m - 2$ and $S \cup \{k\} \in \mathcal{W}$. There are two possibilities:

If $j \notin S$ it is $S \subseteq N \backslash \{j, k\}$ and, since $j \succsim_m k$, $|S| \le m - 2$ and $S \cup \{k\} \in \mathcal{W}$, we have $S \cup \{j\} \in \mathcal{W}$. But it is also true that $S \subseteq N \backslash \{i, j\}$ and, since $i \succsim_m j$, we have $S \cup \{i\} \in \mathcal{W}$.

If $j \in S$, let $S' = S \backslash \{j\}$. Then $S' \cup \{k\} \subseteq N \backslash \{i, j\}$ and $S' \cup \{k\} \cup \{j\} = S \cup \{k\} \in \mathcal{W}$. Since $i \succsim_m j$ and $|S' \cup \{k\}| = |S| \le m - 2$, it is $S' \cup \{k\} \cup \{i\} \in \mathcal{W}$. But $S' \cup \{i\} \subseteq N \backslash \{j, k\}$ and, since $j \succsim_m k$ and $|S' \cup \{i\}| = |S| \le m - 2$, we have $S' \cup \{i\} \cup \{j\} = S \cup \{i\} \in \mathcal{W}$. $\qquad\square$

Proposition 10. *Let v be a simple game and $i, j \in N$. Let m, p be two integers with $1 < m < p < n$. Then,*

$$i \succsim_p j \Rightarrow i \succsim_m j$$
$$i \succ_p j \Rightarrow i \succ_m j$$

Proof:
Assume that $i \succsim_p j$. To prove that $i \succsim_m j$, we only need to see that $|\mathcal{C}_i(h)| \ge |\mathcal{C}_j(h)|$ when $m \le h < p$, because the other cases are immediate consequence of the hypothesis. But, since $\varphi_{ji}(\mathcal{C}_j(h)) \subseteq \mathcal{C}_i(h)$ for any h with $1 \le h < p$, in particular it is $|\mathcal{C}_i(h)| \ge |\mathcal{C}_j(h)|$ when $m \le h < p$.

Assume now that $i \succ_p j$, that is to say, $i \succsim_p j$ and $j \not\succsim_p i$, and we will prove that $j \not\succsim_m i$. The fact that $i \succ_p j$ includes two possibilities:

(a) There is some h with $p \le h \le n$ such that $|\mathcal{C}_i(h)| > |\mathcal{C}_j(h)|$. Since $h \ge p$ implies $h \ge m$, it is clear in this case that $j \not\succsim_m i$.

(b) There is some h with $1 \le h < p$ such that $\varphi_{ij}(\mathcal{C}_i(h)) \not\subseteq \mathcal{C}_j(h)$. This fact directly proves that $j \not\succsim_m i$ if $h < m$. If $m \le h < p$ we will see that $|\mathcal{C}_j(h)| < |\mathcal{C}_i(h)|$ and this also proves that $j \not\succsim_m i$. In effect, notice that $\varphi_{ji}(\mathcal{C}_j(h)) \subsetneq \mathcal{C}_i(h)$, because $i \succsim_p j$ implies $\varphi_{ji}(\mathcal{C}_j(h)) \subseteq \mathcal{C}_i(h)$ and if $\varphi_{ji}(\mathcal{C}_j(h)) = \mathcal{C}_i(h)$ it would be $\varphi_{ij}(\varphi_{ji}(\mathcal{C}_j(h))) = \mathcal{C}_j(h) = \varphi_{ij}(\mathcal{C}_i(h))$ in contradiction with the hypothesis. Thus, $|\varphi_{ji}(\mathcal{C}_j(h))| = |\mathcal{C}_j(h)| < |\mathcal{C}_i(h)|$. $\qquad\square$

The former proposition shows that the m–desirability relations form a chain of preorders on the set N of players. Taking into account Propositions 8 and 10 we can write, for any two elements $i, j \in N$:

$$i \succsim_D j \Leftrightarrow i \succsim_n j \Leftrightarrow i \succsim_{n-1} j \Rightarrow i \succsim_{n-2} j \Rightarrow \cdots \Rightarrow i \succsim_3 j \Rightarrow i \succsim_2 j \Leftrightarrow i \succsim_1 j \Leftrightarrow i \succsim_d j$$
$$i \succ_D j \Leftrightarrow i \succ_n j \Leftrightarrow i \succ_{n-1} j \Rightarrow i \succ_{n-2} j \Rightarrow \cdots \Rightarrow i \succ_3 j \Rightarrow i \succ_2 j \Leftrightarrow i \succ_1 j \Leftrightarrow i \succ_d j$$

Notice, in particular, that all m–desirability relations coincide for $n \leq 3$, and that for $n = 4$ they coincide either with the desirability or with the weak–desirability relation. As a consequence of this fact the m–desirability relations only appear as new preorders for $n \geq 5$.

In the following example all the m-desirability relations are shown.

Example 1. Let $N = \{1, 2, 3, 4, 5, 6, 7, 8, 9\}$ and $\mathcal{W}^m = \{\{1, 2\}, \{3, 4, 5\}, \{6, 7, 8, 9\}\}$

$$
\begin{array}{lll}
1 \approx_D 2 & 3 \approx_D 4 \approx_D 5 & 6 \approx_D 7 \approx_D 8 \approx_D 9 \\
1 \approx_7 2 & 3 \approx_7 4 \approx_7 5 & 6 \approx_7 7 \approx_7 8 \approx_7 9 \\
1 \approx_6 2 & 3 \approx_6 4 \approx_6 5 & 6 \approx_6 7 \approx_6 8 \approx_6 9 \\
1 \approx_5 2 & 3 \approx_5 4 \approx_5 5 & 6 \approx_5 7 \approx_5 8 \approx_5 9 \\
1 \approx_4 2 & 3 \approx_4 4 \approx_4 5 \succ_4 6 \approx_4 7 \approx_4 8 \approx_4 9 \\
1 \approx_3 2 \succ_3 3 \approx_3 4 \approx_3 5 \succ_3 6 \approx_3 7 \approx_3 8 \approx_3 9 \\
1 \approx_d 2 \succ_d 3 \approx_d 4 \approx_d 5 \succ_d 6 \approx_d 7 \approx_d 8 \approx_d 9
\end{array}
$$

For any integer m with $1 \leq m \leq n$ we can define the concept of m–linear game:

Definition 6. *m–linear game*

A simple game v is m–linear whenever the m–desirability relation $\succsim_m$ is complete. The set of m–linear simple games will be denoted by $\mathfrak{L}(m)$.

From the above results it is clear that $\mathfrak{L}(n) = \mathfrak{L}(n-1)$ coincides with the set of linear games, $\mathfrak{L}(1) = \mathfrak{L}(2)$ is the set of weakly linear games, and

$$\mathfrak{L}(n) = \mathfrak{L}(n-1) \subseteq \mathfrak{L}(n-2) \subseteq \cdots \subseteq \mathfrak{L}(3) \subseteq \mathfrak{L}(2) = \mathfrak{L}(1).$$

Thus, a linear game belongs to $\mathfrak{L}(m)$ for all m ($1 \leq m \leq n$), and for any weakly linear (but not linear) game v there exist some m_0 ($2 \leq m_0 < n-1$) such that $v \notin \mathfrak{L}(m)$ for any $m > m_0$ and $v \in \mathfrak{L}(m)$ for any $m \leq m_0$. In the game of Example 1 it is $n = 9$ and $m_0 = 3$. In the next example we show a weakly linear game which does not belong to any other class of m–linear games (in this case it is $n = 5$ and $m_0 = 4$).

Example 2. Let $N = \{1, 2, 3, 4, 5\}$ and let v be the game defined by

$$\mathcal{W}^m = \{\{1, 2\}, \{1, 3\}, \{2, 4\}\}.$$

This game is not linear (does not belong to $\mathfrak{L}(5) = \mathfrak{L}(4)$) because the desirability relation only gives:

$$1 \succ_D 3 \quad \text{and} \quad 2 \succ_D 4 \succ_D 5.$$

Clearly, v is weakly linear (it belongs to $\mathfrak{L}(1) = \mathfrak{L}(2)$) with

$$1 \approx_d 2 \succ_d 3 \approx_d 4 \succ_d 5.$$

But this game it is not 3–linear (it does not belong to $\mathfrak{L}(3)$) because $\{1, 3\} \in \mathcal{W}$ and $\{2, 3\} \notin \mathcal{W}$ implies $2 \not\succsim_3 1$, and, similarly, $\{2, 4\} \in \mathcal{W}$ and $\{1, 4\} \notin \mathcal{W}$ implies $1 \not\succsim_3 2$.

5 Conclusions

In this paper, a chain of classes of simple games is defined. In all of them, the PBC and the SS indices rank players in the same way, i.e., they are ordinally equivalent. The smallest class in the chain is the class of linear games and the largest one is the class of weakly linear games. We think that these classes will have an interesting role in future works. We can mention two open questions:

a) For $n \geq 6$ it is known that all possible hierarchies are achievable in weakly linear games [8] but not all of them are achievable in linear games. Which is the smallest class $\mathfrak{L}(m)$ such that all hierarchies are already achievable in this class?

b) It is known that the Johnston index is ordinally equivalent to PBC and SS indices in linear games but not in weakly linear games. Which is the largest class $\mathfrak{L}(m)$ in which Johnston index is ordinally equivalent to PBC and SS indices?

Acknowledgements. This research was partially supported by grant MTM2015–66818-P(MINECO/FEDER) from the Spanish Ministry of Economy and Competitiveness (MINECO) and from the European Union (FEDER funds).

References

1. Banzhaf, J.F.: Weighted voting doesn't work: a mathematical analysis. Rutgers Law Rev. **19**(2), 317–343 (1965)
2. Carreras, F., Freixas, J.: Some theoretical reasons for using regular semivalues. In: De Swart, H. (ed.) Proceedings of the International Conference on Logic, Game Theory and Social Choice, LGS, Tilburg, The Netherlands, pp. 140–154 (1999)
3. Carreras, F., Freixas, J.: On ordinal equivalence of power measures given by regular semivalues. Math. Soc. Sci. **55**(2), 221–234 (2008)
4. Coleman, J.S.: Control of collectivities and the power of a collectivity to act. In: Lieberman, B. (ed.) Social Choice, pp. 269–300. Gordon and Breach, New York (1971)
5. Diffo Lambo, L., Moulen, J.: Ordinal equivalence of power notions in voting games. Theor. Decis. **53**(4), 313–325 (2002)
6. Felsenthal, D.S., Machover, M.: The Measurament of Voting Power: Theory and Practice, Problems and Paradoxes. Edward Elgar, Cheltenham (1998)
7. Freixas, J., Marciniak, D., Pons, M.: On the ordinal equivalence of the Johnston, Banzhaf and Shapley power indices. Eur. J. Oper. Res. **216**(2), 367–375 (2012)
8. Freixas, J., Pons, M.: Hierarchies achievable in simple games. Theor. Decis. **68**(4), 393–404 (2010)
9. Freixas, J., Zwicker, W.S.: Weighted voting, abstention, and multiple levels of approval. Soc. Choice Welfare **21**(3), 399–431 (2003)
10. Isbell, J.R.: A class of simple games. Duke Math. J. **25**(3), 423–439 (1958)
11. Johnston, R.J.: On the measurement of power: some reactions to Laver. Environ. Plann. A **10**(8), 907–914 (1978)
12. Penrose, L.S.: The elementary statistics of majority voting. J. Roy. Stat. Soc. **109**(1), 53–57 (1946)
13. Roy, S.: The ordinal equivalence of the jhonston index and the established notions of power. In: Econophisics and Economics of Games, Social Choices and Quantitative Techniques, pp. 372–380. Nex Economics Windows, Part II (2010)

14. Shapley, L.S.: A value for n-person games. In: Tucker, A.W., Kuhn, H.W. (eds.) Contributions to the Theory of Games II, pp. 307–317. Princeton University Press, Princeton (1953)
15. Shapley, L.S., Shubik, M.: A method for evaluating the distribution of power in a committee system. Am. Polit. Sci. Rev. **48**(3), 787–792 (1954)
16. Taylor, A.D., Zwicker, W.S.: Simple games: desirability relations, trading, and pseudoweightings. Princeton University Press, New Jersey (1999)

Sub-Coalitional Approach to Values

Izabella Stach[(⊠)]

Faculty of Management, AGH University of Science and Technology,
Krakow, Poland
istach@zarz.agh.edu.pl

Abstract. The behavioral models of classical values (like the Shapley and Banzhaf values) consider the contributions to coalition S as contributions delivered by the players individually joining such a coalition as it is being formed; i.e., $v(S) - v(S \setminus \{i\})$. In this paper, we propose another approach to values where these contributions are considered as given by sets of players: $(v(S) - v(S \setminus R))$, where S, R are subsets of the set of all players involved in cooperative game v. Based on this new approach, several sub-coalitional values are proposed, and some properties of these values are shown.

Keywords: Coalition formation · Cooperative games · Sub-coalitional values · Values

1 Introduction

The bargaining model that leads to some game values consists of the formation of a grand coalition through the successive addition of players to an initially unitary coalition; see, for example, [1]. The value that comes out depends on the way in which coalitions are treated from the point of view of combinatorics (permutations of the value Shapley [2], combinations of the value Banzhaf [3], and so on). These models do not take into account the possibility of joining a pre-constituted group of players in order to form a coalition, as often happens in various situations in politics, finance, and so on. The values considered in this work (we call them sub-coalitional values) admit this possibility. In particular, we discuss some properties of these sub-coalitional values.

This paper is organized as follows. Section 2 presents notations and preliminary definitions that refer to games in characteristic form and values. Section 3 presents definitions of some classical values. Section 4 presents the sub-coalition approach to values. Section 5 is dedicated to concluding remarks and further developments.

2 Preliminaries

In this section, we introduce notations and definitions that refer to cooperative games and values.

A game in characteristic function form represents situations where some players (companies, buyers, sellers, parties, shareholders, countries, and so on) can attain

© Springer International Publishing AG 2017
N.T. Nguyen et al. (Eds.): TCCI XXVII, LNCS 10480, pp. 74–86, 2017.
https://doi.org/10.1007/978-3-319-70647-4_6

common advantages by forming coalitions. Classical market games for economies with private goods are an example of this game. Let $N = \{1, 2,..., n\}$ be a finite set of players. A cooperative game is a pair (N, v) where $v: 2^N \to R$, the characteristic function, is a real-valued function defined on the subsets of N (called coalitions) such that $v(\emptyset) = 0$. Cooperative game v is called superadditive if $v(S \cup T) \geq v(S) + v(T)$ holds for all $S, T \subseteq N$. By G_N, we denote all n-person superadditive cooperative games on N. If, for player $i \in N$, the following condition holds $v(S \cup \{i\}) = v(\{s\})$ for all $S \subseteq N$, then he is called a dummy player. Given two games $v, \cdot w \in G_N$, and two parameters $c, d \geq 0$, their linear combination $(c \cdot v + d \cdot w)$ is defined by $(c \cdot v + d \cdot w)(S) = c \cdot v(S) + d \cdot w(S)$ for all $S \subseteq N$.

A value is a mapping $\psi : G_N \to R^n$ that assigns vector $\psi(v) = (\psi_1(v), \psi_2(v), \ldots, \psi_n(v))$ to each cooperative game v. Component $\psi_i(v)$ is called an individual value of player $i \in N$ in v.

Regarding values, some frequent conditions are imposed. Here, we recall only the following properties:

- Individual rationality. Value ψ fulfils the individual rationality postulate if, for all $i \in N$, the following condition holds: $\psi(\{i\}) \geq v(\{i\})$.
- Efficiency. Value ψ fulfills the *efficiency property* if $\sum_{i \in N} \psi_i(v) = v(N)$ for all $v \in G_N$.

  Generally, this property states that the power of grand coalition N is totally distributed among the players in game v.
- Dummy player. Value ψ satisfies the *dummy player property* if, for each $S \subseteq N \setminus \{i\}$ and all $v \in G_N$, $v(S \cup \{i\}) = v(S) + v(\{i\})$ implies $\psi_i(v) = v(\{i\})$. Informally, this property states that player i (who does not contribute to any coalition) has a measure of power equal to $v(\{i\})$.
- Symmetry. If, for all $v \in G_N$ and for each $i \in N$ and each permutation $\pi : N \to N$ $\psi_i(v) = \psi_{\pi(i)}(\pi(v))$ where $\pi(v)(S) = v(\pi^{-1}(S))$, then we said that value ψ satisfies the *symmetry property*. This property is also called the anonymity property. Informally, this property states that the "symmetric" players should have equal power.
- Linearity. A value satisfies the *linearity property* if, for every games v and w and $c > 0$, the following conditions hold: $\psi(v + w) = \psi(v) + \psi(w)$ and $\psi(cv) = c\psi(v)$.

Let us define the ordered partition of N ($|N| = n$). A *ordered partition* of set N into k parts (i.e., blocks of players or coalitions) where $2 \leq k \leq n$ is any permutation of k nonempty subsets S of N such that every element of N belongs to one and only one of the subsets. Equivalently, permutation $\pi = (S_1,..., S_k)$, is a partition of N if:

(1) no element of $\pi = (S_1,..., S_k)$ is empty (i.e., $S_i \neq \emptyset$ for all $1 \leq i \leq k$);
(2) the union of the elements of π is equal to N (i.e., $S_1 \cup S_2 \cup \ldots \cup S_k = N$);
(3) the intersection of any two elements of π is empty (i.e., the elements of π are pairwise disjoint: $S_i \cap S_j = \emptyset$ for $i \neq j$ and $i, j \in \{1, 2, \ldots, k\}$).

Note that the definition of the ordered partition present above is a bit different than the definition of a partition in the combinatorial theory. Here, we define an ordered partition like a permutation of subsets; and in the combinatorial theory, a partition is any collection of subsets satisfying the three conditions mentioned above; see, for example, [4]. For example, with $N = \{1, 2\}$, we have one partition of N: $\{\{1\}, \{2\}\}$ and two different ordered partitions: $(\{1\}, \{2\})$ and $(\{2\}, \{1\})$.

Let $\Pi_k(N)$ be the set of all ordered partitions of N into k subsets (blocks, coalitions) and $\Pi(N)$ be the set of all possible ordered partitions of N. When the number of players of N is equal to $|N| = n$, then we have $\Pi(N) = \Pi_2(N) \cup \ldots \cup \Pi_n(N)$. Thus, the number of all possible ordered partitions of N is equal to

$$|\Pi(N)| = \sum_{k=2}^{n} |\Pi_k(N)|.$$

Let us note that $|\Pi_k(N)| = k! S(n, k)$, where $S(n, k) = \frac{1}{k!} \sum_{j=0}^{k} (-1)^j \binom{k}{j} (k-j)^n$ is the Stirling number of the second kind (see, for instance, [5]). As for $j = k$ and $n \geq 2 (k-j)^n = 0$, we can write

$$|\Pi_k(N)| = \sum_{j=0}^{k-1} (-1)^j \binom{k}{j} (k-j)^n$$

and

$$|\Pi(N)| = \sum_{k=2}^{n} \sum_{j=0}^{k-1} (-1)^j \binom{k}{j} (k-j)^n.$$

Note that, for $n \geq 2$, the number of all classical partitions of set N is equal to

$$\sum_{k=2}^{n} \frac{1}{k!} \sum_{j=0}^{k-1} (-1)^j \binom{k}{j} (k-j)^n.$$

3   Classical Values

In this section, we recall the definitions of the Banzhaf, Shapley, and Tijs values. The values considered here are based on marginal contributions that are brought by individual players.

Henceforth, if not stated otherwise, all of the games considered are superadditive cooperative games.

3.1   The Banzhaf Value

Let $v \in G_N$ and $i \in N$. The absolute Banzhaf value, β, introduced by Banzhaf [3], is given by:

$$\beta_i(v) = \frac{1}{2^{n-1}} \sum_{S \subseteq N} (v(S) - v(S \backslash \{i\})).$$

The normalized Banzhaf value, β', is define as:

$$\beta'_i(v) = \frac{v(N)}{\sum\limits_{j \in N} \sum\limits_{S \subseteq N} (v(S) - v(S \backslash \{j\}))} \sum_{S \subseteq N} (v(S) - v(S \backslash \{i\})).$$

See also [6].

3.2 The Shapley Value

Let $v \in G_N$ and $i \in N$. The Shapley value is defined as follows:

$$\sigma_i(v) = \sum_{S \subseteq N} \frac{(s-1)!(n-s)!}{n!} (v(S) - v(S \backslash \{i\})$$

where $s = |S|$ is the cardinality of coalition S.

In 1953, Shapley proved that the value given by the formula above is the unique value that satisfies four axioms: efficiency, dummy player, linearity, and symmetry (see [2] and [7]).

3.3 The Tijs Value

Tijs proposed the τ-value in [8] for a particular class of cooperative games called quasi-balanced games. In [9], Driessen and Tijs extended the τ-value to all cooperative games. This extension does not possess all of the properties of the τ-value defined for quasi-balanced games. However, in this paper we consider only the τ-value defined in [8]. Before giving the formula of the Tijs value, we need several notations.

The *upper vector* $b^v = (b^v_1, b^v_2, \ldots, b^v_n) \in R^n$ and *gap function* $g^v : 2^N \to R$ of game $v \in G_N$ are defined, respectively, as:

$$b^v_i = v(N) - v(N \backslash \{i\}) \quad \text{for all } i \in N$$

$$g^v(S) = \sum_{j \in S} b^v_j - v(S) \quad \text{for all } S \subseteq N.$$

Each component b^v_i of the upper vector is called the marginal contribution of player i, with respect to grand coalition N. Note that $b^v_i \geq v(\{i\})$ for all $i \in N$ and superadditive game v.

Let $\rho^v_i = \min\limits_{S \ni i} g^v(S)$ for all $i \in N$; then, vector $\rho^v = (\rho^v_1, \rho^v_2, \ldots, \rho^v_n) \in R^n$ is called the *concession vector*. The class of quasi-balanced n-person games, Q^v_B, is defined as:

$$Q_B^v = \{v \in G_N : g^v(S) \geq 0 \text{ for all } S \subseteq N \text{ and } \sum_{j \in N} \rho_j^v \geq g^v(N)\}.$$

The Tijs value (τ-value) for any quasi-balanced game $v \in Q_B^v$ is defined as:

$$\tau(v) = \begin{cases} b^v & \text{if} \quad g^v(N) = 0 \\ b^v - \frac{g^v(N)}{\sum\limits_{j \in N} \rho_j^v} \rho^v & \text{if} \quad g^v(N) > 0 \end{cases}$$

When $g^v(N) = 0$, each player i receives his marginal contribution; while, if $g^v(N) > 0$ (which equivalently means that it is impossible for each player to receive at least his marginal contribution), then each player i forgoes his share:

$$\frac{g^v(N)}{\sum\limits_{j \in N} \rho_j^v} \rho_i^v.$$

In 1987, Tijs axiomatized his value; see [10]. For more about τ-value, see also [11].

4 New Approach to Values

In cooperative games, when the grand coalition N will be formed, the problem arises with the sharing of total winnings $v(N)$ among the players.

Among the models able to represent such a payoff distribution over the players, the first was the Shapley value. The concept of "value" suggested by Shapley in [2] resulted in a breakthrough in the theory of cooperative games, as earlier attempts to solve cooperative games did not ensure the existence and uniqueness of the solution. The Shapley value abandoned the classical methods based on dominance in favor of a suitable model of bargaining founded in a system of axioms (efficiency, symmetry, dummy player, and linearity).

The Shapley model assumes that construction of grand coalition N is realized through the successive addition of players, one by one, in a randomly chosen order to the coalition that is being formed. Each player i gets his marginal contribution which he contributes to the coalition S; i.e., amount $v(S) - v(S\setminus\{i\})$, opportunely weighted. Subsequently, Shapley's fourth axiom becomes a point of discussion, and this encouraged many authors to invent other values (based on diverse axiomatic assumptions and/or bargaining models) that are able to describe other types of real situations with a common structure:

$$\psi_i(v) = \alpha \sum_{S \subseteq N} \beta(v(S) - v(S\setminus\{i\}))$$

where α and β are opportune parameters of the combinatorial character depending on the order in which the player joined S; see Table 1.

Table 1. Coefficients α and β in formulae of classical values

Value	α	β
The Shapley value	$1/n$	$1 \Big/ \binom{n-1}{s-1}$
The absolute Banzhaf value	$1/2^{n-1}$	1
The normalized Banzhaf value	$\dfrac{v(N)}{\sum\limits_{j\in N}\sum\limits_{S\subseteq N}(v(S)-v(S\setminus\{j\}))}$	1

The previous model of values considers contributions to coalition S during its formation as supplied by players individually joined to such a coalition: $v(S) - v(S\setminus\{i\})$. In this work, we suggest an approach to the values based on sub-coalition, where these contributions are considered as given by sets R of players: $v(S) - v(S\setminus R)$. Such as "pictorial" example of the construction of a mosaic or puzzle that can be realized by the successive addition of singular pieces or just-constructed blocks of the whole picture; see Figs. 1, 2, and 3.

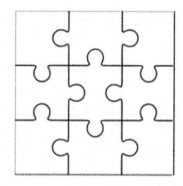

Fig. 1. Grand coalition formed

Fig. 2. Formation of grand coalition by successive addition of singular players

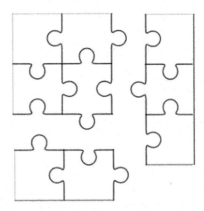

Fig. 3. Formation of grand coalition by successive addition of blocks of players

A Leading Example

Let us consider a three-person cooperative game in the following characteristic form:

$$v(\emptyset) = 0, \ v(\{1\}) = 1, \ v(\{2\}) = v(\{3\}) = 0$$
$$v(\{1, 2\}) = 4, \ v(\{1, 3\}) = 3, \ v(\{2, 3\}) = 2$$
$$v(\{1, 2, 3\}) = 5.$$

In the case of the Shapley value, the negotiation takes place as follows. Initially, Player 2 will appeal to Player 1 and ask him to join his coalition. The contribution that "2" brings to the new coalition is, thus, calculated. If "1" plays alone and wins $v(\{1\}) = 1$, while the coalition recently formed between 1 and 2 wins $v(\{1, 2\}) = 4$; therefore, the contribution of Player 2 is equal to $v(\{1, 2\}) - v(\{1\}) = 4 - 1 = 3$. If Player 3 now offers to join the coalition already formed $\{1, 2\}$, the contribution that he brings to the new coalition is: $v(\{1, 2, 3\}) - v(\{1, 2\}) = 5 - 4 = 1$. This situation is represented in the first row of Table 2. The sequence of additions, however, could take place otherwise for example, Player 3 is first added to 1, then 2 is added to $\{1, 3\}$; in this case, the distribution is shown in the second row of Table 2.

Let us consider all of the possible sequences of the grand coalition formation as equally probable. Then, for each player, the Shapley value can be calculated as his/her expected payoff corresponding to the contribution s/he made to the various coalitions multiplied by the probability that s/he is in a "pivotal" position, which is being added to an already-formed coalition. In our case, this probability is 1/6 for each player; so, the Shapley value distributes the overall payoff as follows: 15/6, 9/6, and 1 to Players 1, 2, and 3, respectively.

As indicated above, the proposed model describes situations where the formation of the grand coalition takes place with additions not only of individual players but also of pre-established groups, in their turn formed with the same mechanism. Based on this approach, we propose new values that we will call the *SC-values* (Sub-coalition marginal Contributions values).

Table 2. Shapley value calculation

Formation of grand coalition		Marginal contribution of players to coalitions		
		1	**2**	**3**
1	{1}{2}{3}	1	3	1
2	{1}{3}{2}	1	2	2
3	{2}{1}{3}	4	0	1
4	{2}{3}{1}	3	0	2
5	{3}{1}{2}	3	2	0
6	{3}{2}{1}	3	2	0
Totals		15	9	6
The Shapley value		$2.5 = 15/6$	$1.5 = 9/6$	$1 = 6/6$

A big problem is how to divide marginal contributions among the members of sub-coalitions R. For the games with only two players, there is no problem, as the possible sub-coalitions are sets of singular players. The problem instead arises for games with more than two players.

This topic of sharing marginal contributions to the coalitions formed by players joining in random order in cooperative games was just touched by Malawski in [12]. He introduced the so-called procedural values, considering different procedures under which the players can only share their marginal contributions with their predecessors in the ordering. All procedural values satisfy efficiency, symmetry, and linearity properties. Furthermore, it has been shown that, along with two monotonicity postulates, these properties characterize the class of procedural values.

The various method of sharing the marginal contribution among the members of a sub-coalition once applied leads to different sub-coalitional values. In order to share a sub-coalitional marginal contribution, we have to consider a subgame composed of the members of sub-coalition.

4.1 SC-Values Base on the Ordered Partitions of N

If the order in which the players join a coalition is important (like in the Shapley value) and we decide to divide the marginal contribution of members of a sub-coalition equally, we receive an SC-value that we will call the *egalitarian SC-Shapley value*. If any player takes his marginal contribution in each coalition that he belongs, and the rest (which is never negative in the superadditive cooperative games) is divided equally, we obtain the SC-value that we call the *Harsanyi-Nash SC-Shapley value*.

Let us consider equiprobable all of the grand coalition formation sequences. In the leading example, the first six options (shown in the first six rows of Table 3) reproduce exactly what were seen for the value of Shapley (see Table 2). The other options are added by a new bargaining method.

Table 3 shows how many different possibilities there are to form grand coalition N in the leading example. In a general case, if the number of players of N is equal to $|N| = n$, the number of possibilities to form a grand coalition is equal to the number of

Table 3. Scheme of calculation of egalitarian SC-Shapley value in leading example

Formation of grand coalition		Marginal contribution of players to coalitions		
		1	**2**	**3**
1	{1}{2}{3}	1	3	1
2	{1}{3}{2}	1	2	2
3	{2}{1}{3}	4	0	1
4	{2}{3}{1}	3	0	2
5	{3}{1}{2}	3	2	0
6	{3}{2}{1}	3	2	0
7	{1}{2, 3}	1	2	2
8	{2, 3}{1}	3	1	1
9	{2}{1, 3}	2.5	0	2.5
10	{1, 3}{2}	1.5	2	1.5
11	{3}{1, 2}	2.5	2.5	0
12	{1, 2}{3}	2	2	1
Totals		27.5	18.5	14
The egalitarian SC-Shapley value		$\frac{55}{24} \approx 2.29$	$\frac{37}{24} \approx 1.54$	$\frac{28}{24} \approx 1.17$

all possible ordered partitions of N. Thus, in the leading example, we have $|\Pi(N)| =$

$$\sum_{k=2}^{3} \sum_{j=0}^{k-1} (-1)^j \binom{k}{j} (k-j)^3 = 12 \text{ possible formations of grand coalition } N = \{1, 2, 3\};$$

see Sect. 2.

The SC-values defined in this way (i.e., the egalitarian SC-Shapley and Harsanyi-Nash SC-Shapley values) satisfy the symmetry and efficiency properties and do not satisfy the dummy player condition; see the example of the game given in Table 4. Furthermore, in comparing with the Shapley value, the individual rationality condition is lost in the case of the egalitarian SC-Shapley value and preserved by the SC-Harsanyi-Nash value; see the example of the game in Table 5.

Table 4. SC-values and classical values in example of game with dummy player 3

Game: $v(\{1\}) = v(\{1,3\}) = 1$, $v(\{2\}) = v(\{3\}) = 0$, $v(\{2, 3\}) = 0$, $v(\{1, 2\}) = v(N) = 4$	Players power		
	1	**2**	**3**
The Shapley, two Banzhaf, and Tijs values	2.500	1.500	0.000
The egalitarian SC-Shapley value	2.208	1.458	0.333
The Harsanyi-Nash SC-Shapley value	2.375	1.375	0.250
The normalized SC-Banzhaf value	2.011	1.474	0.514
The SC-Banzhaf value	3.000	1.875	0.875
The SC-Tijs value	2.250	1.375	0.375

Table 5. SC-values and classical values in example of game with dictator player 1

Game: $v(\{1\}) = v(\{1,2\}) = v(\{1,3\}) = v(N) = 1,$ $v(\{2\}) = v(\{3\}) = v(\{2,3\}) = 0$	Players power		
	1	2	3
σ, β, β', τ, the Harsanyi-Nash SC-Shapley, normalized Banzhaf value, and SC-Tijs values	1	0	0
The egalitarian SC-Shapley value	10/12	1/12	1/12
The SC-Banzhaf value	0.75	0	0

4.2 SC-Values Base on the Partitions of N

In case the order of pre-constituted sub-coalitions is not important in the formation of a grand coalition, we propose the SC-Banzhaf and SC-Tijs values. For these values, the number of possible formation of the grand coalition is equal to the number of all possible partitions of grand coalition N. So, taking into account the leading example where we have three players ($n = 3$), the number of all possible constructions of the grand coalition is equal to $\sum_{k=2}^{n} \frac{1}{k!} \sum_{j=0}^{k-1} (-1)^j \binom{k}{j} (k-j)^n = 4$; see Table 6 and Sect. 2.

Let us first introduce the SC-Banzhaf and normalized SC-Banzhaf values. The *normalized SC-Banzhaf value* is defined as follows. In the first step, the marginal contribution of sub-coalition R is calculated simply from the following formula: $v(N) - v(N\setminus R)$. In Step 2, if R is not a singular player, the marginal contribution is shared among the members following the Banzhaf value and characteristic function of original game v. In Step 3, we sum up all marginal contributions for each player and opportunely weigh the results in order to obtain the efficient value. More precisely, let us consider Case 3 in Table 6. Here, the marginal contribution of Player 2 is equal to $v(N) - v(\{1,3\}) = 5 - 3 = 2$ and the marginal contribution of sub-coalition $\{1,3\}$ is equal to: $v(N) - v(\{2\}) = 5$. Now, we have to share 5 among Players 1 and 3. So we have a two-person subgame v' with $v'(N') = 5$, grand coalition $N' = \{1,3\}$, and $v'(S) = v(S)$ for all $S \subset N'$. To share the winnings of this subgame between Players 1 and 3, we calculate the players' marginal contributions to this sub-coalition using the characteristic function of the original game (in our case, $v'(N') - v(N'\setminus\{i\}) = 5 - v(\{1,3\}\setminus\{i\})$ for $i = 1, 3$) and then we opportunely weighted the result (i.e. multiplying by the normalization factor (K) which is obtain as division of $v'(N')$ by the sum of contributions for all members of v'; in our case $K = 5/9$) and received the following split of winnings: 25/9, 20/9 for Players 1 and 3, respectively; see Table 6.

Analogously, we define the *SC-Banzhaf value*. The difference lies in the sharing of the marginal contributions of sub-coalitions R. Here, we applied the Banzhaf idea in the sub-games, so the weighted coefficient is equal to $2^{|N'|}$. In the case of the SC-Banzhaf value and the leading example, we obtain the following distribution of power: 2.75, 1.75, and 1.25 for Players 1, 2, and 3, respectively. By the way, this is the same distribution as the Banzhaf value given in this example.

Generally, the *SC-Tijs value* is also similarly defined as the SC-Banzhaf values. However, we share the winnings of the sub-coalitions in the subgames following the

Table 6. Scheme of calculation of normalized SC-Banzhaf value in leading example

Formation of grand coalition		Marginal contribution of players to coalitions		
		1	2	3
1	{1}{2}{3}	3	2	1
2	{1}{2, 3}	3	2	2
3	{2}{1, 3}	25/9	2	20/9
4	{3}{1, 2}	25/9	20/9	1
Totals		104/9	74/9	56/9
Normalized SC-Banzhaf value		$\frac{260}{117} \approx 2.22$	$\frac{185}{117} \approx 1.58$	$\frac{140}{117} \approx 1.20$

Tijs value and original characteristic game. More precisely; in the first step, we calculate the Tijs value for each of the possible cases of forming grand coalition N. However, note that we have to treat sub-coalitions R as singular players in the cases of forming a grand coalition like that of the 2nd, 3rd, and 4th row in Table 6. Then, in the second step in each subgame v', we again apply the Tijs value in order to share the marginal contribution of the sub-coalition. Note that the characteristic function of a subgame is defined as follows: grand coalition $N' = R$, $v'(N')$ is equal to the marginal contribution of R (i.e., the Tijs value of R assigned in the previous step) and $v'(S) = v(S)$ for all $S \subset R = N'$. Finally, we sum up the results obtained for each player and properly normalized to $v(N)$; i.e., we divide them by the number of all possible formations of grand coalition N. Taking into account the leading example as well as the Tijs and SC-Tijs values, we receive the following distributions of power for Players 1, 2, and 3: (8/3, 5/3, 2/3) and (119/48, 71/48, 50/48), respectively.

The SC-values based on combinations like the normalized SC-Banzhaf and SC-Tijs values, thanks to their construction, preserve the symmetry and efficiency properties. However, they lose the dummy player property; see the example of the game given in Table 4. The SC Banzhaf value is not efficient and does not satisfy the individual rationality (see the example in Table 5) nor the dummy properties (see the example in Table 4), but it fulfills the symmetry property.

5   Concluding Remarks

In general, the bargaining models of the classical values consider the contributions to the coalition in the formation as supplied by players individually joined to such a coalition: $(v(S) - v(S \setminus \{i\}))$. In this work, we proposed an approach to values based on sub-coalitions where these contributions are considered as given by sets of players: $(v(S) - v(S \setminus R))$. With this approach, we proposed several sub-coalitional values like the egalitarian SC-Shapley, Harsanyi-Nash SC-Shapley, SC-Banzhaf, and Tijs values (see Sect. 4). Of course, this approach opens up many problems; in particular, the determination of the formula for the new modified values and their axiomatic derivations. Such problems will be the subject of a forthcoming article.

A further development is to apply the sub-coalitional bargaining model to other values like the Public Good value [13] or Public Help value [14], for example.

Very relevant are the models taking into account the coalition structure: situations where not all coalitions are equally probable (see [15]). Several authors have applied suitable modifications to values taking this notion into account. Amongst the more well-known articles, we should mention the work undertaken by Owen in [16], Derks and Peters in [17], Amer and Carreras in [18], and Malawski in [19]. One further development is to consider the SC-values in the context of games with a priori unions.

One of the possible applications of sub-coalitional values is to calculate the power of the players in situations like those described in [20]. This means to calculate the power of the countries in the European Union when the countries are aggregated in groups: any group around one lead country.

Finally, another interesting development of this research regards the connection between the results already achieved by Grabisch and Roubens [21]. They introduced the interaction index that has several interesting points in common with the proposal presented in this paper.

Acknowledgements. Research is financed by the statutory funds (no. 11/11.200.322) of the AGH University of Science and Technology. The author expresses gratitude to Gianfranco Gambarelli and Cesarino Bertini for useful suggestions and precious help.

The author also gratefully acknowledge the helpful comments and suggestions of two referees.

References

1. Gambarelli, G.: Value of a game. In: Dowding, K. (ed.) Encyclopedia of Power, pp. 683–684. SAGE Publications, Los Angeles (2011)
2. Shapley, L.S.: A value for n-person games. In: Tucker, A.W., Kuhn, H.W. (eds.) Contributions to the Theory of Games II, pp. 307–317. Princeton University Press, Princeton (1953)
3. Banzhaf, J.F.: Weighted voting doesn't work: a mathematical analysis. Rutgers Law Rev. **19**, 317–343 (1965)
4. Lucas, J.F.: Introduction to abstract mathematics. Ardsley House Publishers Inc., New York (1990)
5. Abramowitz, M., Stegun, I.A. (eds.): Stirling numbers of the second kind. In: Handbook of Mathematical Functions with Formulas, Graphs, and Mathematical Tables, pp. 824–825. Dover, New York (1972)
6. Bertini, C., Stach, I.: Banzhaf voting power measure. In: Dowding, K. (ed.) Encyclopedia of Power, pp. 54–55. SAGE Publications, Los Angeles (2011)
7. Bertini, C.: Shapley value. In: Dowding, K. (ed.) Encyclopedia of power, pp. 600–603. SAGE Publications, Los Angeles (2011)
8. Tijs, S.H.: Bounds for the core and the τ-value. In: Moeschlin, O., Pallaschke, D. (eds.) Game Theory and Mathematical Economics, pp. 123–132. North Holland, Amsterdam (1981)
9. Driessen, T.S.H., Tijs, S.H.: Extensions and modifications of the τ-value for cooperative games. In: Hammer, G., Pallaschke, D. (eds.) Selected Topics in Operations Research and Mathematical Economics. Lecture Notes in Economics and Mathematical Systems, vol. 226, pp. 252–261. Springer, Heidelberg (1984). https://doi.org/10.1007/978-3-642-45567-4_18

10. Tijs, S.H.: An axiomatization of the τ-value. Math. Soc. Sci. **13**, 177–181 (1987)
11. Stach, I.: Tijs value. In: Dowding, K. (ed.) Encyclopedia of Power, pp. 667–670. SAGE Publications, Los Angeles (2011)
12. Malawski, M.: "Procedural" values for cooperative games. Int. J. Game Theor. **42**(1), 305–324 (2013)
13. Holler, M.J., Li, X.: From public good index to public value. an axiomatic approach and generalization. Control Cybern. **24**(3), 257–270 (1995)
14. Bertini, C., Stach, I.: On public values and power indices. Decis. Making Manuf. Serv. **9**(1), 9–25 (2015)
15. Aumann, R.J., Drèze, J.: Cooperative games with coalition structures. Int. J. Game Theor. **3**, 217–237 (1974)
16. Owen, G.: Values of games with a priori unions. In: Henn, R., Moeschlin, O. (eds.) Mathematical Economics and Game Theory. Lecture Notes in Economics and Mathematical Systems, vol. 141, pp. 76–88. Springer, Heidelberg (1977). https://doi.org/10.1007/978-3-642-45494-3_7
17. Derks, J., Peters, H.: A Shapley value for games with restricted coalitions. Int. J. Game Theor. **21**, 351–360 (1993)
18. Amer, R., Carreras, F.: Cooperation indices and weighted Shapley values. Math. Oper. Res. **22**, 955–968 (1997)
19. Malawski, M.: Counting power indices for games with a priori unions. Theor. Decis. **56**, 125–140 (2004)
20. Mercik, J.W., Ramsey, D.: The effect of Brexit on the balance of power in the European Union Council: An approach based on pre-coalitions. In: Nguyen, N.T., et al. (eds.) TCCI XXVII, LNCS Transactions on Computational Collective Intelligence, vol. 10480, pp. 87–107. Springer, Cham (2017)
21. Grabisch, M., Roubens, M.: An axiomatic approach to the concept of interaction among players in cooperative games. Int. J. Game Theor. **28**, 547–565 (1999)

The Effect of Brexit on the Balance of Power in the European Union Council: An Approach Based on Pre-coalitions

Jacek Mercik[1]([✉]) and David M. Ramsey[2]

[1] WSB University in Wrocław, Wrocław, Poland
jacek.mercik@wsb.wroclaw.pl
[2] Department of Operations Research, Wrocław University of Science and Technology, Wrocław, Poland
david.ramsey@pwr.edu.pl

Abstract. This article anlayses the change in the balance of power in the European Union Council due to the United Kingdom leaving (referred to as Brexit). This analysis is based on the concept of power indices in voting games where natural coalitions, called pre-coalitions, occur between various players (or parties). The pre-coalitions in these games are assumed to be formed around the six largest member states (after Brexit, the five largest), where each of the remaining member states joins the pre-coalition based around the large member state which is the most similar according to the subject of the vote. This is illustrated by an example. We consider adaptations of three classical indices: the Shapley-Shubik, Banzhaf-Penrose and Johnston indices based on the concept of a consistent share function (also called quotient index). This approach can be interpreted as a two-level process of distributing power. At the upper level, power is distributed amongst pre-coalitions. At the lower level, power is distributed amongst the members of each pre-coalition. One of the conclusions of the research is that removing the UK from the voting game means that the power indices of small countries actually decrease. This seems somewhat surprising as the voting procedure in the EU council was designed to be robust to changes in the number and size of member states. This conclusion does not correspond to a general result, but does indicate the difficulty of defining voting rules which are robust to changes in the set of players.

Keywords: Voting games · Power indices · Coalitions · European Union Council

1 Introduction

Shapley and Shubik (1954) were the first to explicitly define what they called a power index in a voting game (a form of cooperative game). The power of player (voter) i is measured by the probability that, when the players enter a coalition in random order, player i swings the coalition from being a losing

© Springer International Publishing AG 2017
N.T. Nguyen et al. (Eds.): TCCI XXVII, LNCS 10480, pp. 87–107, 2017.
https://doi.org/10.1007/978-3-319-70647-4_7

coalition to a winning coalition. Banzhaf proposed a power index based on the number of vulnerable coalitions in which player i is critical (i.e. when player i leaves such a coalition, it changes from a winning coalition to a losing one). It was later recognized that the derivation of this index is identical to the analysis of voting systems presented by Penrose (1946), although he did not call his solution a power index. The so-called Banzhaf-Penrose index was then adapted by Johnston (1978) to take into account the fact that the smaller the number of critical players belonging to a vulnerable coalition, the greater the power wielded by each critical player.

One problem with these classical measures of power is that they do not take into account the characteristics of players, i.e. they assume that players do not have preferences regarding who they form coalitions with. Myerson (1977) and Owen (1986) consider models in which the voters are positioned on a graph and form coalitions with neighbours before forming coalitions with more distant players.

Another possible approach is to consider voting games in which there exist natural pre-coalitions of similar voters. This is a natural approach when there are a reasonably large number of players (parties) involved in a political vote. In such a case, the existence of such pre-coalitions would greatly reduce the negotiating costs. Owen presented adaptions of the Shapley-Shubik value (Owen 1977) and of the Banzhaf-Penrose value (Owen 1982) for games in which natural coalitions exist. These results were generalised by Alonso-Meijide and Fiestras-Janeiro (2002) and the monotonicity properties of such indices were considered by Alonso-Meijide et al. (2009). In that paper, the authors used the concept of a quotient index, based on the idea of consistency introduced by Van den Brink and Van der Laan (2001), and note that the indices defined by Owen are of this form. The values of such indices can be calculated by splitting the full game into two levels. The upper level is the voting game played between the pre-coalitions, in which the weight of a pre-coalition is given by the sum of the weights of the members of the pre-coalition. The lower level is given by a set of induced subgames played between the members of a pre-coalition. The power indices for the games played at both levels can be calculated using classical methods. The solution of the game played at the upper level measures the power of the pre-coalitions. By taking the appropriate weighted average of the vector of power indices in the relevant subgames played within a pre-coalition, we can measure the power of a player within a pre-coalition. The power index of a player is then given by the product of the power index of its pre-coalition and the power index describing the influence of a player within its own pre-coalition. We use the indices proposed by Owen (1977, 1982), together with the appropriately adapted version of the Johnston index (Johnston 1978).

It might seem that such an approach would complicate the calculation of power indices. However, the time required to solve a game is exponential in the number of players. Hence, by solving the game in this way, the calculation time is much reduced, since the games considered involve significantly fewer players than the full game. We apply this approach to solving the voting game played by the EU council according to the Treaty of Lisbon rules, both before and after Brexit (games with 28 and 27 players, respectively). The power indices were derived by full enumeration using a program written by the authors in the R package. In addition, we used the "ipnice" program available at www.warwick. ac.uk/~ecaae which calculates the Banzhaf index for voting games based on two weight vectors. This algorithm calculates the power indices using generating functions, which greatly speeds up the calculation (see Leech 2003). The results obtained using this algorithm agreed with the results we obtained for the Banzhaf value based on full enumeration.

Since a very powerful player is leaving the EU, a political union with huge economic and social influence, it is interesting to those researching in the political, social and economic sciences to understand how the balance of power within the union will change. In this article, we look in particular at the effect that the UK leaving the EU will have on decision making with regard to the economic sphere. Similar analyses could be made with regard to other issues, such as agriculture or the environment, given data describing the appropriate factors.

Due to the varying population sizes of the member states and the nature of the voting procedures adopted in the EU, such procedures have been a natural source for researchers as a tool for developing fair voting systems when players have naturally different weights (Bertini et al. 2005 and Turnovec 2009). Mercik et al. (2004) consider the interaction of national and partisan (party political) interests in such voting procedures. Malawski (2004) considers the allocation of payoffs (power) in the framework of simple games, which strictly include the set of majority voting games. For a more general overview of power indices and their application to voting systems see Gambarelli and Stach (2009) and Bertini et al. (2013).

It should also be noted that as a result of the workshop on "Quantitative Methods of Group Decision Making", to which this book is dedicated, the authors discovered that Gianfranco Gambarelli and Izabella Stach (2017) have been working independently on an approach to defining power indices of co-operative games with given pre-coalitions (appearing also in this volume).

The article is arranged as follows: Sect. 2 describes the rules of voting in the EU council as prescribed by the Treaty of Lisbon, together with the three classical measures of power which will be applied here. Section 3 describes how the pre-coalitions are defined. The derivation of the power indices of the pre-coalitions in the voting game (the external index) is considered in Sect. 4. Section 5 presents the derivation of the power index for individual players within a pre-coalition (the internal index). The relation between the approach presented here and the general concept of quotient indices is given in Sect. 6. Numerical results for the voting game played by the members of the EU council are presented in Sect. 7. The final section gives some conclusions and directions for future research.

2   Voting Procedure in the EU Council and Classical Power Indices

This analysis is based on the rules introduced by the Lisbon Treaty in order to reflect possible changes in the membership of the EU and the populations of individual states (see Kóczy 2012 and http://www.consilium.europa.eu/en/council-eu/voting-system/). In order for a vote to be passed, the following two conditions must be satisfied:

1. 55% of the member states must be in favour (i.e. at present, 16 of 28 states, after Brexit, 15 of 27 states). It should be noted that in special cases 72% of the countries must be in favour, but this variant is not considered here.
2. The states voting for a proposition should represent at least 65% of the EU population, with the additional condition that to block a proposition any coalition representing at least 35% of the population must contain at least four states.

The present members of the EU and their percentage share of the EU population (both pre- and post-Brexit) are given in Table 1. These percentages are calculated based on the populations as of 1st Jan, 2015 according to Eurostat (http://ec.europa.eu/eurostat/data/database).

In terms of the set of winning coalitions, the condition that at least four states are required to block a proposition only relates to the six largest states, since any coalition of three countries representing at least 35% of the EU population must contain three of these states. These six states (Germany, France, the

Table 1. Member states of the EU and their relative population sizes (pre- and post-Brexit)

Country	% Pop. (pre)	% Pop. (post)	Country	% Pop. (pre)	% Pop. (post)
1. Germany	15.96	18.29	15. Austria	1.69	1.94
2. France	13.06	14.97	16. Bulgaria	1.42	1.63
3. UK	12.74	n.a	17. Denmark	1.11	1.27
4. Italy	11.96	13.71	18. Finland	1.08	1.24
5. Spain	9.14	10.47	19. Slovakia	1.07	1.23
6. Poland	7.48	8.57	20. Ireland	0.91	1.04
7. Romania	3.91	4.48	21. Croatia	0.83	0.95
8. Netherlands	3.33	3.82	22. Lithuania	0.57	0.65
9. Belgium	2.22	2.54	23. Slovenia	0.41	0.47
10. Greece	2.13	2.44	24. Latvia	0.39	0.45
11. Czech Rep.	2.07	2.37	25. Estonia	0.26	0.30
12. Portugal	2.04	2.34	26. Cyprus	0.17	0.19
13. Hungary	1.94	2.22	27. Luxembourg	0.11	0.13
14. Sweden	1.92	2.20	28. Malta	0.08	0.09

United Kingdom, Italy, Spain and Poland) are significantly larger (according to population) than the others. In addition, the only such coalition that does not contain Germany is the coalition formed by France, the United Kingdom and Italy, the second to fourth largest countries. Hence, this condition seems to be aimed at reducing the power of the six largest nations, especially Germany.

We now consider three classical power indices. Let N be the set of all players (countries) and S denote the set of players in a coalition. Also, we denote the total number of players by n and the number of players in coalition S by s. The characteristic function v is defined on the space of coalitions, such that $v(S) = 1$ when coalition S is winning and otherwise $v(S) = 0$. A winning coalition S is said to be vulnerable, if there exists at least one member of that coalition who can turn it into a losing coalition by leaving S. Such a member is called critical.

We consider the following three power indices:

1. The Shapley-Shubik index:

$$\phi_i(v) = \sum_{S \subseteq N \setminus \{i\}} \frac{s!(n-s-1)!}{n!} [v(S \cup \{i\}) - v(S)], \qquad (1)$$

where $\phi_i(v)$ is the Shapley-Shubik index for player i based on the voting game with characteristic function v (Shapley and Shubik 1954). In intuitive terms, this is the probability that by entering a coalition, it is player i who turns it from a losing coalition to a winning coalition when players enter coalitions in a random order. Thus, it is a measure of the power to implement a decision.

2. The normalized Banzhaf-Penrose (BP) index (Penrose 1946 and Banzhaf 1965) is obtained by dividing the number of coalitions for which player i is critical by the sum of these numbers over the set of players. Hence, the sum of the normalized values is 1, as is the sum of the Shapley-Shubik indices. If $\beta_i(v)$ is the number of coalitions for which i is critical and $\overline{\beta}_i(v)$ is the normalized BP index, then

$$\beta_i(v) = \sum_{S \subseteq N \setminus \{i\}} v(S \cup \{i\}) - v(S); \quad \overline{\beta}_i(v) = \frac{\beta_i(v)}{\sum_{i=1}^{n} \beta_i(v)}. \qquad (2)$$

The normalized BP index is interpreted as the power of a player to block a decision.

3. The Johnston power index (Johnston 1978) is an adaptation of the Banzhaf index. It is obtained by weighting each coalition for which player i is critical by the reciprocal of the number of players who are critical in that coalition. As such, it remains a measure of the power to block a decision, while giving more weight to players who are more likely to be either the only player able to block a decision or one of a small number of such members. If $\gamma_i(v)$ is the non-normalized Johnston power index and $\overline{\gamma}_i(v)$ is the normalized value, then

$$\gamma_i(v) = \sum_{S \subseteq N \setminus \{i\}} \frac{v(S \cup \{i\}) - v(S)}{c(S \cup \{i\})}; \quad \overline{\gamma}_i(v) = \frac{\gamma_i(v)}{\sum_{i=1}^{n} \gamma_i(v)}, \qquad (3)$$

where $c(S)$ denotes the number of critical players in the coalition S. It should be noted that here we sum over the set of coalitions in which player i is critical. It follows that for such sets $c(S) \geq 1$ and hence the sum is well-defined.

Note that these indices are various weighted sums based on the set of coalitions for which a given player (country) is critical. This fact is crucial to the method described in Sect. 5 to derive how power is split within a pre-coalition.

3   Defining Pre-coalitions

When the number of players is large, unless the negotiation process has some form of structure, then it seems likely that the costs of negotiation would increase more rapidly than the benefits to be obtained by each player being able to voice their own opinion. This is illustrated by the concept of *liberum veto* used in the Polish-Lithuanian parliament in the 17th Century, which lead to institutional paralysation (Roháč 2008) and, on the other hand, the common practice of party discipline in modern parliaments. In the case of the EU under its present regulations, it seems reasonable to assume that smaller countries would seek to align themselves with larger countries who have similar goals to them, since otherwise they could not achieve the amount of support according to population in order to pass favourable legislation. On the other hand, larger states need the support of a large number of states, so naturally look to build alliances with smaller countries with similar aims. For this reason, we consider a model in which pre-coalitions are formed around the six largest member states (Germany, France, the UK, Italy, Spain and Poland) according to the similarity of countries' economic situations and then these pre-coalitions are assumed to negotiate to form larger coalitions. These six countries are chosen for the following reasons:

1. They are significantly larger than any of the other EU member states.
2. They represent a wide variety of viewpoints, as well as economic, historic and political positions, within the EU.

The economic factors considered are as follows:

1. GDP growth (as a percentage).
2. GDP per capita (in terms of purchasing power).
3. Debt (as a percentage of GDP).
4. Trade surplus (as a percentage of GDP).
5. Inflation (as a percentage).
6. Unemployment rate.
7. Consumption (in terms of purchasing power).
8. Classification of individual consumption by purpose (COICOP). This is an indicator of the structure of consumption and is negatively correlated with wealth.

It should be noted that this choice of variables is subjective (one could, for example, add government spending as a proportion of GDP). However, the data are used for illustrative purposes and the resulting partition of member states according to their economic situation seems reasonable and reflects many of the natural inclinations of countries to align themselves (in particular, the tendency of the post-communist countries to group together and the close ties between France and Germany, although the ties between these two countries are more multi-dimensional). Of course, such an approach is only appropriate for motions which are predominantly economic in nature. When considering reforms to the Common Agricultural Policy, one should use data relevant to agriculture (e.g. percentage employed in agriculture, percentage of GDP produced by agriculture, level of EU agricultural subsidies). Similar approaches could be defined for motions related to e.g. the environment or immigration policy.

Since these variables are not all measured in the same units, they are individually standardized by subtracting the relevant mean and then dividing by the standard deviation. The Euclidean distance between two countries according to these standardized data was used as a measure of dissimilarity (the matrix of these distances is given as an appendix). As stated above, it was assumed that the pre-coalitions formed around the six largest countries (after Brexit, the five largest countries remaining). The smaller countries joined the pre-coalition based around the country which was economically most similar to it. This is a somewhat simplistic approach and in the future we wish to investigate probabilistic models where the probability that smaller countries join a particular pre-coalition depends on the distance to the country around which that pre-coalition is formed. In the voting game played pre-Brexit, the six pre-coalitions formed according to this approach, together with their weights according to the voting procedure, are given in Table 2. Table 3 gives the weights of the pre-coalitions in the voting game played post-Brexit. For convenience, each pre-coalition will be referred to according to the largest country in the pre-coalition. It should be noted that based on these pre-coalitions, the voting rule which states that at least four states are required to block a rule will no longer have any effect, since any coalition including three of the six largest states will also include other states.

4 Determining the Power Indices of the Pre-coalitions

In order to determine the appropriate power index of a pre-coalition, we consider the voting games defined when the pre-coalitions are treated as single players. Each of the indices can be calculated by determining the set of vulnerable coalitions, and calculating the appropriate sum over this set. Tables 4 and 5 give the set of vulnerable coalitions and the corresponding sets of critical players for the pre-Brexit and post-Brexit games, respectively. In addition, these tables present the number of permutations corresponding to a given pre-coalition turning a coalition from being losing to being winning (i.e. being the "swing" party) when pre-coalitions come together in a random order. It should be noted that each

Table 2. Pre-coalitions in the pre-Brexit voting game along with their weights. Post-Brexit, Ireland joins the pre-coalition formed around France.

	Pre-coalition 1	Pre-coalition 2	Pre-coalition 3	Pre-coalition 4
	Germany	France	United Kingdom	Italy
	Sweden	Netherlands	Ireland	Portugal
	Austria	Belgium		Cyprus
	Denmark	Finland		
	Luxembourg			
No. of countries	5	4	2	3
% of EU population	20.79	19.69	13.65	14.10
	Pre-coalition 5	Pre-coalition 6		
	Spain	Poland		
	Greece	Romania	Lithuania	
	Croatia	Czech Rep.	Slovenia	
		Hungary	Latvia	
		Bulgaria	Estonia	
		Slovakia	Malta	
No. of countries	3	11		
% of EU population	12.10	19.60		

Table 3. The weights of the pre-coalitions in the post-Brexit game.

Largest country	Germany	France	Italy	Spain	Poland
No. of countries	5	5	3	3	11
% of EU population	23.83	23.61	16.24	13.86	22.46

possible permutation of the pre-coalitions defines a well defined combination of vulnerable coalition (the first winning coalition formed) and swing player (who is a critical player in that coalition). For example, when the Polish pre-coalition is the swing player in the pre-Brexit game and the coalition formed is {D, F, UK, Pol}, Poland must be the fourth pre-coalition in the permutation, any of the six possible orderings of Germany, France and the UK in the first three positions is possible, combined with the two possible orderings of Italy and Spain in the final two positions. Hence, there are 12 permutations corresponding to the vulnerable coalition {D, F, UK, Pol} where Poland is the swing player.

The power indices for these two games can be directly calculated based on the information given in Tables 4 and 5. In order to calculate the Shapley-Shubik value, we calculate the number of permutations in which a pre-coalition is the swing player by summing the number of permutations for which a pre-coalition is critical and then dividing by the total number of permutations of the pre-coalitions, $n!$, where n is the number of pre-coalitions. The BP index is calculated by counting the number of coalitions in which a given pre-coalition is critical and dividing it by the sum of the number of critical players over

Table 4. Vulnerable coalitions in the pre-Brexit game and the set of critical pre-coalitions in each case: D - Germany, F - France, UK - United Kingdom, It - Italy, Es - Spain, Pol - Poland

Vulnerable coalition	Critical players	Permutations/critical player
{D, F, UK, Pol}	{D, F, UK, Pol}	12
{D, F, It, Pol}	{D, F, It, Pol}	12
{D, UK, It, Pol}	{D, UK, It, Pol}	12
{F, UK, It, Pol}	{F, UK, It, Pol}	12
{D, F, Es Pol}	{D, F, Es, Pol}	12
{F, UK, Es, Pol}	{F, UK, Es, Pol}	12
{D, It, Es, Pol}	{D, It, Es, Pol}	12
{F, It, Es, Pol}	{F, It, Es, Pol}	12
{D, F, UK, It, Es}	{D, F, UK, It, Es}	24
{D, F, UK, It, Pol}	{Pol}	24
{D, F, UK, Es Pol}	{Pol}	24
{D, F, It, Es, Pol}	{Pol}	24
{D, UK, It, Es, Pol}	{D, Pol}	24
{F, UK, It, Es, Pol}	{F, Pol}	24

Table 5. Vulnerable coalitions in the post-Brexit game and the set of critical pre-coalitions in each case.

Vulnerable coalition	Critical players	Permutations/critical player
{D, F, Pol}	{D, F, Pol}	4
{D, F, It, Es}	{D, F, It, Es}	6
{D, F, It, Pol}	{D, F, It, Pol}	6
{D, F, Es, Pol}	{D, F, Es, Pol}	6
{D, It, Es, Pol}	{D, It, Es, Pol}	6
{F, It, Es, Pol}	{F, It, Es, Pol}	6

all the vulnerable coalitions. The Johnston index is obtained by dividing the sum of the reciprocals of the number of critical players over the set of coalitions in which a given pre-coalition is critical by the total number of vulnerable coalitions. These indices are given in Tables 6 and 7 for the pre-Brexit and post-Brexit games, respectively.

It should be noted that the Polish pre-coalition is particularly powerful in the pre-Brexit game. This is due to the fact that any coalition of two pre-coalitions which includes Poland is a blocking coalition (since the number of countries involved is always large enough to block a motion) and the only other blocking coalition of two pre-coalitions is formed by Germany and France (which represents more than 35% of the EU population).

Table 6. Power indices of the pre-coalitions in the pre-Brexit game.

	Germany	France	UK	Italy	Spain	Poland
Shapley	10/60	10/60	7/60	7/60	7/60	19/60
BP	4/24	4/24	3/24	3/24	3/24	7/24
Johnston	44/300	44/300	29/300	29/300	29/300	125/300

Table 7. Power indices of the pre-coalitions in the post-Brexit game.

	Germany	France	Italy	Spain	Poland
Shapley	14/60	14/60	9/60	9/60	14/60
BP	5/21	5/21	3/21	3/21	5/21
Johnston	2/8	2/8	1/8	1/8	2/8

5   Determining the Power of a Member State Within a Pre-coalition

In order to determine the power of each member state in these two games, it is necessary to determine the power of a member state within a pre-coalition. This is done by considering subgames corresponding to each vulnerable coalition. Suppose pre-coalition i is critical in a given coalition. Then in order to calculate the power, in the corresponding subgame, of a member state within a pre-coalition, we assume that all the members of all the other pre-coalitions in the vulnerable coalition vote for a motion, while all the members of the pre-coalitions outside of the vulnerable coalition vote against the motion. This assumption leads to an induced subgame played by the members of pre-coalition i in which a specified number of member states must vote for the motion and these states must represent a defined proportion of the EU population. For example, in the post-Brexit game, the German pre-coalition can play the five subgames described in Table 8 corresponding to the five coalitions in which the German pre-coalition is critical. These are described in Table 8. For example, in the subgame corresponding to the coalition {D, F, Pol}, it is assumed that all the countries in the French and Polish pre-coalitions vote for the proposition (a total of 16 countries representing 46.07% of the EU population), while the countries in the Italian and Spanish pre-coalitions are assumed to vote against the proposition. It follows that the member states in the German pre-coalition play a subgame in which in order to pass a proposition, it is simply necessary that the countries in the German pre-coalition voting for the proposition must represent 18.93% of the EU population (so that the 65% threshold is exceeded). The power indices for these subgames can be calculated using standard methods (e.g. using the approach described in the previous section).

In order to calculate the power indices of the member states within pre-coalition i, it is necessary to calculate the appropriate weighted average of the indices derived for the subgames played by pre-coalition i. In the case of calcu-

Table 8. Power indices of the pre-coalitions in the post-Brexit game.

Vulnerable coalition	No. of countries required	% of EU population required	Weight
{D, F, Pol}	0	18.93	4
{D, F, It, Es}	4	11.29	6
{D, F, Es, Pol}	0	5.07	6
{D, It, Es, Pol}	0	12.44	6
{D, F, It, Pol}	0	2.69	6

lating either the BP index or the Johnston index, the vulnerable coalitions are equally weighted. Hence, to calculate the appropriate power index for a member state within pre-coalition i, we simply average the power indices over the appropriate subgames. In the case of the Shapley-Shubik index, the weight of a subgame is assumed to be proportional to the number of permutations of pre-coalitions corresponding to that subgame, i.e. the permutations corresponding to pre-coalition i being a swing player (the derivation of these weights was described in the previous section). Hence, since the sum of the weights must be equal to one, the weights of the first subgame described above is $1/7$ and the weights of the remaining subgames are all equal to $3/14$.

The power index of a member state in the game as a whole is obtained by multiplying the appropriate power of that member state within its pre-coalition (the internal index) by the same power index of the pre-coalition in the game with defined pre-coalitions (the external index). Any such index satisfies the efficiency condition, since the sum of the values of any power index over the member states is by definition equal to one. For example, the Shapley-Shubik index for Germany in the five games listed above are $3/4$, $2/5$, $3/4$, 1 and $1/2$, respectively. It follows that the Shapley-Shubik index for Germany within its pre-coalition is given by

$$v_D^{i,S} = \frac{1}{7} \times \frac{3}{4} + \frac{3}{14}\left(\frac{2}{5} + \frac{3}{4} + 1 + \frac{1}{2}\right) = \frac{27}{40},$$

where the superscripts indicate that this is the internal Shapley power index and the subscript indicates the country. Hence, the Shapley-Shubik index for Germany in the full game with pre-coalitions is given by

$$v_D^S = v_D^{e,S} v_D^{i,S} = \frac{7}{30} \times \frac{27}{40} = \frac{63}{400}.$$

6 Relation to the General Concept of Quotient Indices

We consider here the relation of the indices considered above to the indices introduced by Owen (1977, 1977) for cooperative games played by the set of players N with defined pre-coalitions $C = \{C_1, C_2, \ldots C_k\}$, where $\cup_{i=1}^{k} C_i = N$

and $K = \{1, 2, \ldots k\}$. Based on Alonso-Meijide *et al.* (2009) and assuming that player i, $1 \le i \le n$, belongs to coalition C_j, $1 \le j \le k$, the form of an index from the set of quotient indices can be written in the following form:

$$g_i(v, C) = \sum_{R \subseteq C \backslash C_j} p^u_{R,j} \sum_{T \subseteq C_j \backslash i} p^\ell_{T,i}[v(Q_R \cup T \cup i) - v(Q_R \cup T)], \qquad (4)$$

where $g_i(v, C)$ is the power index for player i in the voting game with value function v when the set of pre-coalitions is given by C, R denotes a coalition of pre-coalitions, Q_R denotes the set of players in such a coalition, T is a set of players within pre-coalition C_j, $p^u_{R,j}$ is an appropriately defined weight at the upper (external) level, i.e. the voting game played between pre-coalitions and $p^\ell_{T,i}$ is an appropriately defined weight at the lower (internal) level, i.e. the voting game played within a pre-coalition. According to this sum, the power of an individual player comes from the scenarios in which that player is a critical player in its pre-coalition in an appropriately defined internal voting game given that the pre-coalition is critical in the external voting game. Due to the fact that the weights can be split as a quotient into internal and external terms, the internal sum in Eq. (4) can be interpreted as a power index in the internal subgame. In the case of the index based on the Shapley-Shubik index, (see Owen 1977)

$$p^\ell_{T,i} = \frac{t!(c_j - t - 1)!}{c_j!}, \quad p^u_{R,j} = \frac{r!(k - r - 1)!}{k!},$$

where t is the number of players in T, c_j is the number of players in pre-coalition C_j, r is the number of pre-coalitions in the set R of pre-coalitions. It follows that the internal sum defines the Shapley-Shubik index of player i in an appropriately defined subgame given that the coalition C_j is critical. Note that the sum of these indices for each such subgame is by definition one. Hence, it follows that

$$\sum_{i \in C_j} g_i(v, C) \equiv g_{C_j}(v, C) = \sum_{R \subseteq C \backslash C_j} p^u_{R,j}[v(Q_R \cup C_j) - v(Q_R)], \qquad (5)$$

where $g_{C_j}(v, C)$ is the power of pre-coaliton C_j in the voting game. It thus follows that the external index defined in Sect. 4 can be interpreted as the Shapley-Shubik of the pre-coalition in the external game.

Finally, it should be noted that the overall index weights the coalitions in which the pre-coalition C_j is critical according to the number of permutations where the pre-coalition C_j is critical. It follows that to define the appropriate power index of a player within a pre-coalition, the corresponding weighting of the powers in the subgames must be carried out, as described in the previous section. In conclusion, the overall power index for a player corresponding to the Shapley-Shubik index can be interpreted as the product of the two indices describing internal and external power.

With regard to the index based on the BP index, the story is slightly different. Based on Alonso-Meijide (2009), the internal and external weights are given by $p^\ell_{T,i} = 2^{1-k}$ and $p^u_{R,j} = 2^{1-c_j}$. In this case, the indices are not normalised (i.e. they do not sum to one), but at each stage of the summation, the coalitions are equally weighted. It follows that the index defined in this paper is an appropriately normalised version of the index proposed by Owen (1982). Arguing in a similar way, the Johnston index is an appropriately normalised version of the original index defined for voting games without pre-coalitions. In the case of each of these three indices, when the pre-coalitions are individual players, these indices reduce to the standard normalised indices.

7   Numerical Results

A program was written in the R package to evaluate power indices by complete enumeration for a voting game based on two weight vectors. One problem with this approach is that the complexity of this calculation is exponential in the number of players (there are 2^n possible coalitions). For example, the complete enumeration of the three classical indices for the pre-Brexit game (with 28 players) took around 36 h on a laptop (4 MB RAM, Intel Core i3), while the indices for games with 11 players (i.e. the greatest number of players in a subgame considered here) were calculated essentially immediately. For this reason, although programs to calculate power indices based on voting games with pre-coalitions are more complex to write, the computing time required to calculate the appropriate power indices is hugely reduced. Table 9 gives the values for the pre-Brexit voting game. The first value is the index calculated according to the classical index. The value in brackets corresponds to the voting game based on the pre-coalitions described in Sect. 3. Table 10 gives the corresponding values for the post-Brexit game.

Comparing the power of the countries according to the classical indices, each of them indicates that the power of small countries is relatively large in proportion to their share of the EU population, particularly the Banzhaf index. This is in agreement with the observations of Alonso-Meijide et al. (2009). The Johnston index ascribes a particularly high power to Germany, the largest country. It should be noted that the difference between the populations of Germany and France (the second largest country) is greater than the population of the majority of member states. As a result, Germany will be a critical player when there are a relatively small number of critical players more often than any of the other countries and this is reflected in the Johnston index.

Comparing the values of the classical indices pre-Brexit and post-Brexit, it is surprising that the power of the smallest countries decreases due to the UK leaving (although this fall is only marginal). This might be explained by the fact that after Brexit, according to the Lisbon treaty, the support of only 15 of the 27 member states is required to pass a motion (a lower proportion than pre-Brexit). The stipulation requiring support of member states representing 65% of the population of the member states remains unchanged. Hence, the size of a country seems to be becomes relatively more important in comparison to the voting weight resulting from simply being a member state. This is particularly visible in the case of the Johnston index, where the values of the index of the 19

Table 9. Power indices for the pre-Brexit voting game scaled to sum to 100. The first value is the classical index. The value in brackets is based on the pre-coalitions described in Sect. 3

Country	Pre-coaltion	% pop.	Shapley-Shubik	BP	Johnston
1. Germany	Germany	15.96	14.43 (11.83)	10.21 (11.46)	21.96 (10.41)
2. France	France	13.06	11.25 (9.44)	8.46 (9.76)	12.34 (8.86)
3. UK	UK	12.74	10.93 (9.17)	8.27 (10.42)	11.71 (8.06)
4. Italy	Italy	11.96	10.17 (7.22)	7.85 (8.19)	10.61 (6.44)
5. Spain	Spain	9.14	7.53 (5.56)	6.21 (6.25)	7.16 (4.83)
6. Poland	Poland	7.48	6.32 (6.65)	5.08 (6.12)	5.04 (11.51)
7. Romania	Poland	3.91	3.75 (4.74)	3.79 (4.19)	2.94 (6.61)
8. Netherlands	France	3.33	3.28 (2.78)	3.48 (2.67)	2.49 (2.22)
9. Belgium	France	2.22	2.42 (2.78)	2.90 (2.67)	1.74 (2.22)
10. Greece	Spain	2.13	2.36 (3.89)	2.86 (4.17)	1.69 (3.22)
11. Czech Rep.	Poland	2.07	2.31 (3.42)	2.82 (3.31)	1.65 (4.30)
12. Portugal	Italy	2.04	2.29 (3.06)	2.81 (3.19)	1.64 (2.42)
13. Hungary	Poland	1.94	2.21 (3.35)	2.76 (3.22)	1.59 (4.14)
14. Sweden	Germany	1.92	2.20 (1.83)	2.75 (2.05)	1.58 (1.70)
15. Austria	Germany	1.69	2.03 (1.83)	2.63 (2.05)	1.48 (1.70)
16. Bulgaria	Poland	1.42	1.83 (2.81)	2.49 (2.79)	1.37 (3.49)
17. Denmark	Germany	1.11	1.60 (0.58)	2.33 (0.56)	1.25 (0.43)
18. Finland	France	1.08	1.58 (1.67)	2.31 (1.56)	1.24 (1.38)
19. Slovakia	Poland	1.07	1.58 (2.35)	2.31 (2.43)	1.24 (3.00)
20. Ireland	UK	0.91	1.46 (2.50)	2.22 (2.08)	1.18 (1.61)
21. Croatia	Spain	0.83	1.40 (2.22)	2.18 (2.08)	1.16 (1.61)
22. Lithuania	Poland	0.57	1.21 (1.95)	2.05 (1.77)	1.07 (2.15)
23. Slovenia	Poland	0.41	1.10 (1.69)	1.96 (1.48)	1.03 (1.77)
24. Latvia	Poland	0.39	1.09 (1.68)	1.95 (1.47)	1.02 (1.75)
25. Estonia	Poland	0.26	1.00 (1.59)	1.89 (1.36)	0.99 (1.67)
26. Cyprus	Italy	0.17	0.93 (1.39)	1.84 (1.11)	0.96 (0.81)
27. Luxembourg	Germany	0.11	0.89 (0.58)	1.81 (0.56)	0.94 (0.43)
28. Malta	Poland	0.08	0.87 (1.44)	1.79 (1.03)	0.94 (1.30)

smallest countries actually fall as a result of Brexit. In general, how the removal of a player affects the power indices of the least powerful players will depend on the exact form of the voting game and the player who leaves. The power of the smallest countries can increase or decrease, as can be seen from the following game: Suppose there are 4 states, whose populations are in the ratio 2:2:2:1. Consider a voting game where a winning coalition must contain the majority of the states and represent at least 65% of the total population. It is simple to show that any coalition of three states is a minimal winning coalition. Hence, any normalized index describing the power of the smallest state must be equal to 1/4. However, when one of the large states is removed from the game, the small state is never critical (i.e. has zero power). However, if only 55% of the total population need to be represented, then the game is reduced to one in which

Table 10. Power indices for the post-Brexit voting game scaled to sum to 100. The first value is the classical index. The value in brackets is based on the pre-coalitions described in Sect. 3

Country	Pre-coalition	% pop.	Shapley-Shubik	BP	Johnston
1. Germany	Germany	18.29	17.32 (15.75)	11.92 (14.34)	24.45 (18.16)
2. France	France	14.97	13.28 (13.36)	9.98 (12.39)	15.41 (14.31)
3. Italy	Italy	13.71	12.03 (13.33)	9.18 (12.38)	13.05 (11.11)
4. Spain	Spain	10.47	9.00 (8.33)	7.66 (7.30)	9.38 (7.64)
5. Poland	Poland	8.57	6.99 (7.43)	6.54 (7.47)	7.32 (11.08)
6. Romania	Poland	4.48	4.00 (3.70)	4.01 (3.74)	3.11 (4.28)
7. Netherlands	France	3.82	3.52 (4.75)	3.68 (5.61)	2.62 (5.64)
8. Belgium	France	2.54	2.60 (2.94)	3.02 (3.15)	1.74 (2.91)
9. Greece	Spain	2.44	2.53 (3.33)	2.97 (3.49)	1.68 (2.43)
10. Czech Rep.	Poland	2.37	2.47 (2.44)	2.93 (2.79)	1.64 (2.49)
11. Portugal	Italy	2.34	2.45 (0.83)	2.91 (0.95)	1.63 (0.69)
12. Hungary	Poland	2.22	2.37 (2.36)	2.85 (2.67)	1.57 (2.34)
13. Sweden	Germany	2.20	2.35 (2.28)	2.84 (2.75)	1.56 (2.03)
14. Austria	Germany	1.94	2.17 (2.28)	2.70 (2.75)	1.45 (2.03)
15. Bulgaria	Poland	1.63	1.94 (1.90)	2.54 (2.08)	1.33 (1.70)
16. Denmark	Germany	1.27	1.68 (2.28)	2.35 (2.75)	1.19 (2.03)
17. Finland	France	1.24	1.66 (1.28)	2.33 (1.57)	1.18 (1.24)
18. Slovakia	Poland	1.23	1.65 (1.50)	2.32 (1.65)	1.17 (1.25)
19. Ireland	France	1.04	1.52 (1.00)	2.22 (1.09)	1.11 (0.91)
20. Croatia	Spain	0.95	1.45 (3.33)	2.17 (3.49)	1.08 (2.43)
21. Lithuania	Poland	0.65	1.24 (1.01)	2.01 (1.02)	0.99 (0.59)
22. Slovenia	Poland	0.47	1.11 (0.89)	1.92 (0.77)	0.94 (0.40)
23. Latvia	Poland	0.45	1.09 (0.86)	1.91 (0.74)	0.94 (0.39)
24. Estonia	Poland	0.30	0.98 (0.72)	1.82 (0.59)	0.89 (0.31)
25. Cyprus	Italy	0.19	0.91 (0.83)	1.76 (0.95)	0.87 (0.69)
26. Luxembourg	Germany	0.13	0.86 (0.75)	1.73 (0.84)	0.85 (0.75)
27. Malta	Poland	0.09	0.87 (0.52)	1.71 (0.30)	0.84 (0.17)

only a simple majority is required (both before and after the removal of a large state). In this case, any normalized power index for the small state must increase from $1/4$ to $1/3$.

Now we compare the classical indices with those obtained on the basis of pre-coalitions. It is clear that the power of a particular member state depends on both the power of the pre-coalition that it is a member of and its position within that pre-coalition. For example, in the pre-Brexit game the pre-coalition formed around Poland has a very strong position, since it can form a blocking coalition with any other pre-coalition. The only other pair of pre-coalitions that can form a blocking coalition are the ones formed around Germany and France. As a result of this, the values of the Shapley-Shubik index and, in particular, the Johnston index of all the members of the Polish pre-coalition (mostly small countries) are clearly greater for the model based on pre-coalitions. There does not seem to be

Table 11. Economic distance between member states

	2	3	4	5	6	7	8	9	10	11	12	13	14	15	16	17	18	19	20	21	22	23	24	25	26	27	28
1. Germany	2.76	2.90	3.22	5.35	3.99	5.27	1.59	2.43	7.40	3.08	4.41	3.71	2.15	1.78	5.11	1.92	2.32	3.74	4.19	4.73	4.22	3.72	3.79	3.51	4.17	3.85	3.89
2. France	-	1.66	1.36	3.20	3.26	5.03	1.58	1.13	5.03	3.48	2.19	2.98	3.19	2.22	4.51	2.09	1.27	2.72	3.90	3.06	4.08	2.60	3.31	3.97	3.46	5.29	3.90
3. UK	-	-	2.67	4.08	3.72	5.80	1.74	1.95	5.96	3.76	3.38	3.62	3.23	2.37	5.09	2.03	1.91	3.27	3.52	4.19	5.07	3.05	4.15	4.66	4.21	4.86	3.79
4. Italy	-	-	-	3.21	3.52	5.03	2.46	1.61	4.51	3.77	1.71	2.90	3.84	2.72	4.61	3.11	2.23	3.08	4.31	2.66	4.10	2.67	3.44	4.17	3.13	6.05	4.22
5. Spain	-	-	-	-	3.76	5.52	4.25	4.07	3.23	4.93	3.32	4.10	5.20	5.16	4.41	4.69	3.73	3.36	4.41	2.50	4.86	3.06	4.20	5.45	3.20	6.99	4.46
6. Poland	-	-	-	-	-	2.61	3.27	3.83	5.94	2.32	3.51	1.71	4.02	4.49	1.57	3.71	3.28	0.86	4.03	2.54	2.38	1.34	1.93	2.81	3.27	6.28	2.03
7. Romania	-	-	-	-	-	-	4.91	5.26	7.37	3.25	5.25	2.87	5.10	5.89	2.43	5.30	5.09	2.93	5.61	3.86	1.90	3.66	2.52	2.96	4.98	7.38	3.69
8. Netherlands	-	-	-	-	-	-	-	1.49	6.46	3.25	3.36	2.97	1.91	1.50	4.51	0.93	1.31	2.79	3.50	3.80	4.04	2.89	3.10	3.33	3.90	4.06	3.44
9. Belgium	-	-	-	-	-	-	-	-	5.83	3.39	2.39	3.13	1.44	1.44	5.12	2.10	1.89	3.24	3.92	3.67	4.46	3.29	3.49	4.03	4.21	5.05	4.20
10. Greece	-	-	-	-	-	-	-	-	-	7.41	4.36	6.09	7.84	7.01	6.31	6.86	5.63	5.75	7.09	4.22	6.57	5.06	6.48	7.52	4.54	9.50	6.97
11. Czech Rep.	-	-	-	-	-	-	-	-	-	-	3.91	1.72	2.50	3.50	3.28	3.19	3.49	2.03	3.59	3.59	3.28	2.82	1.67	2.16	4.49	5.32	2.33
12. Portugal	-	-	-	-	-	-	-	-	-	-	-	2.68	4.46	3.68	4.84	3.96	3.25	2.93	4.63	2.45	4.24	3.01	3.24	4.40	4.16	7.04	4.48
13. Hungary	-	-	-	-	-	-	-	-	-	-	-	-	3.60	3.76	2.64	3.54	3.30	1.40	4.05	2.30	2.87	1.93	1.34	2.45	3.85	6.36	2.70
14. Sweden	-	-	-	-	-	-	-	-	-	-	-	-	-	2.39	5.16	2.23	3.08	3.51	3.11	4.76	4.64	4.01	3.41	1.67	4.10	6.16	3.11
15. Austria	-	-	-	-	-	-	-	-	-	-	-	-	-	-	5.72	1.78	2.25	3.98	4.35	4.62	5.02	4.06	3.98	4.04	4.84	4.48	4.63
16. Bulgaria	-	-	-	-	-	-	-	-	-	-	-	-	-	-	-	4.80	4.34	2.17	5.25	2.86	2.37	2.38	2.47	3.00	3.68	7.22	3.10
17. Denmark	-	-	-	-	-	-	-	-	-	-	-	-	-	-	-	-	1.36	3.27	4.10	4.28	4.35	3.42	3.50	3.56	3.44	5.99	3.88
18. Finland	-	-	-	-	-	-	-	-	-	-	-	-	-	-	-	-	-	2.90	4.30	3.44	3.88	2.73	3.36	3.56	3.34	4.77	3.98
19. Slovakia	-	-	-	-	-	-	-	-	-	-	-	-	-	-	-	-	-	-	3.64	2.24	2.70	1.39	1.57	2.77	3.43	5.99	2.17
20. Ireland	-	-	-	-	-	-	-	-	-	-	-	-	-	-	-	-	-	-	-	4.81	5.61	3.71	4.43	5.40	4.83	4.50	2.63
21. Croatia	-	-	-	-	-	-	-	-	-	-	-	-	-	-	-	-	-	-	-	-	3.32	2.11	2.43	3.62	2.99	7.18	3.91
22. Lithuania	-	-	-	-	-	-	-	-	-	-	-	-	-	-	-	-	-	-	-	-	-	3.08	2.39	2.38	3.75	6.71	3.79
23. Slovenia	-	-	-	-	-	-	-	-	-	-	-	-	-	-	-	-	-	-	-	-	-	-	2.54	3.41	2.39	6.09	2.31
24. Latvia	-	-	-	-	-	-	-	-	-	-	-	-	-	-	-	-	-	-	-	-	-	-	-	1.67	4.10	4.58	3.11
25. Estonia	-	-	-	-	-	-	-	-	-	-	-	-	-	-	-	-	-	-	-	-	-	-	-	-	4.58	6.00	3.97
26. Cyprus	-	-	-	-	-	-	-	-	-	-	-	-	-	-	-	-	-	-	-	-	-	-	-	-	-	6.45	3.85
27. Luxembourg	-	-	-	-	-	-	-	-	-	-	-	-	-	-	-	-	-	-	-	-	-	-	-	-	-	-	5.43

any systematic tendency of the pre-coalition approach to increase or decrease the values of the Shapley-Shubik and Johnston power indices of small countries, since in the post-Brexit game the pre-coalition formed around Poland no longer has such a strong position relative to the other coalitions and the values of the power indices in this case are more comparable to the classical values. On the other hand, the fact that the values of the BP index for the smallest countries in the pre-coalition formed around Poland are lower than according to the classical value seems to indicate that the introduction of pre-coalitions seems to generally decrease the value of the BP index for small countries (hence making these values more similar to the Shapley-Shubik and Johnston values). Considering small countries from other pre-coalitions seems to support this conclusion, although there are some exceptions (see below).

Even taking into consideration the bargaining position of a pre-coalition, the bargaining position of a country within a pre-coalition can have a significant impact on the value of its power indices in the full voting game. When there are a relatively large number of players, since power indices involve averaging over a very large number of possible coalitions, the power of a country will be very strongly associated with its size. However, when only a small number of countries play the subgames which define the distribution of power within a pre-coalition, then it is very possible that one country can have essentially the same bargaining position as a clearly smaller country. This results from the specific form of the subgames played. Denmark seems to be a very good example of this phenomenon. In the pre-Brexit game, when considering the winning coalitions in all of the possible subgames played by the pre-coalition formed around Germany, the positions of Denmark and Luxembourg are interchangeable, even though Denmark has a population ten times the size of Luxembourg. Hence, the values of Denmark's power indices in these games are very low. However, in the post-Brexit game, Denmark's position in the subgames played by the pre-coalition formed around Germany is interchangeable with the position of Austria, or even Sweden, whose population is nearly twice the size. Hence, the values of Denmark's power indices in the post-Brexit game are greater than would be expected from its size. Portugal is another very good example of this effect. However, in this case, the values of Portugal's power indices in the post-Brexit game are smaller than would be expected from its size.

Note that the monotonicity of values is preserved when comparing the power of pre-coalitions (a pre-coalition cannot have a smaller power index than another pre-coalition containing at most the same number of countries and representing at most the same population) and within pre-coalitions (a country cannot have a smaller power index than another member of the same pre-coalition with a smaller population). However, when comparing countries from different pre-coalitions, monotonicity is not preserved (as noted by Alonso-Meijide et al. 2009). For example, in the pre-Brexit game, the values of Denmark's power indices are exceeded by all of the other countries except for Luxembourg. On the other hand, in the post-Brexit game, the values of Denmark's power indices are significantly greater than those of Portugal, a country with a population more than twice the size of Denmark's.

8   Conclusion and Future Research

This article has described a method of deriving power indices in voting games with a given set of pre-coalitions by adapting the definition of three classical power indices: the Shapley-Shubik, Banzhaf-Penrose and Johnston indices. These concepts are illustrated by the voting game played by the EU Council and defined in the Treaty of Lisbon, where member states are the players and pre-coalitions are groups of similar states. The values of the power indices of the pre-coalitions (the external index) is derived by treating the voting game as a game played by these pre-coalitions, where the weight of a pre-coalition is given by the sum of the weights of the member states in that pre-coalition. The value of each of the power indices for a member state within a pre-coalition is derived by appropriate summing over the set of coalitions in which that pre-coalition is critical.

The value of a power index of a player within in pre-coalition i (the internal index) is given by a weighted sum of indices based on the subgames induced by the set of coalitions in which pre-coalition i is critical. In each of these induced subgames, it is assumed that the other pre-coalitions within the vulnerable coalition vote for the motion and the pre-coalitions outside of the vulnerable coalition vote against the motion. This induces a voting game played within pre-coalition i. In the framework of the EU Council, in such a subgame a defined number of member states from pre-coalition i representing a given proportion of the EU population must support the bill, in order for the bill to be passed.

The value of the power index of an individual player is defined to be the product of the power index of a player within a pre-coalition (the internal index) and the power of the pre-coalition in the full game (the external index). It follows directly that such an index satisfies the property of efficiency (i.e. the sum of the power indices of the individual players is equal to one). However, such an index does not satisfy the condition of monotonicity, i.e. when player j has a greater voting weight than player k, then the value of the power index of player k may be greater. It would be interesting to check what other desirable properties of power indices are (or are not) satisfied. For example, the indices defined here possess the following symmetry properties: (i) if two players with the same voting weights belong to the same pre-coalition, the values of their power indices are equal, (ii) interchanging two players with the same voting weights who belong to different pre-coalitions will simply lead to interchanging the values of the power indices of these two players (in both cases without changing the power indices of the remaining players).

In terms of the effect of Brexit on the balance of power in the EU council, on the basis of the three classical power indices considered, it is somewhat surprising that the power of the smallest countries falls. This seems to be due to the fact that, according to the rules defined in the Lisbon treaty, a smaller percentage of the individual member states is required to pass a motion, while the percentage of the EU population that these member states must represent remains the same. Hence, population size becomes more important relatively to simply being a member state. However, as argued above, this is certainly not a general result for such games.

It is interesting to note that in the pre-Brexit game a strong pre-coalition based around Poland (which is composed almost entirely of relatively new members of the EU in Central Europe) would be a very strong player in the EU council. In practice, such a large coalition would not be a monolith. Also, France and Germany are relatively close in economic terms and tend to co-operate strongly in EU politics (see Kauppi and Widgrén 2007). The model considered here assumes that a member state joins the pre-coalition formed around the large member state which is most similar to it. Also, these pre-coalitions are equally likely to cooperate with any of the other pre-coalitions. A more realistic model would assume that member states are more likely, but not guaranteed, to join the pre-coalition based around the most similar large member state and that pre-coalitions are more likely to enlarge themselves by forming coalitions with pre-coalitions which are most similar to them. Such assumptions would indicate that France is a more powerful player than found here, since it occupies a central position, both economically and politically.

Since the UK is a somewhat isolated (or independent) player in the EU council (both in practice and in accordance with our model), Brexit is unlikely to lead to any major rearrangement of natural coalitions within the EU. On the other hand, the fact that such a large country is leaving the EU might have a significant impact on the balance of power within the EU council. According to the model presented here, the pre-coalition formed around Poland will lose power, since France (or Germany) can now form a blocking coalition with any other pre-coalition. However, since France and Germany have tended to co-operate together, it is likely that Poland's power in the pre-Brexit game is exaggerated.

Although these EU voting rules were devised to take into account the possibility that the number of member states will change, it is particularly interesting to note that, even so, given changes in the membership can have rather systematic effects on the balance of power. In the case of Brexit, it is unsurprising that the power indices of the largest countries increase (since there are less players), but it is surprising that the power of the smallest countries seems to fall. These changes are subtle and depend on the precise nature of the voting rules and which players leave (or join). In the case of the EU, one of the rules states that at least 55% of the member states should favour a motion. However, in practice, the percentage of states that need to support a bill will depend on the number of member states. For example, before Brexit this percentage was 57.14% (16 out of 28), after Brexit, it is only 55.56% (15 out of 27). This indicates that more research is required to investigate voting rules which are robust to changes in membership, while still being practical to use.

Although the UK will leave the EU, it is quite possible that there will eventually be an agreement that the UK will participate in the European free market (in a similar manner to Norway and Switzerland). Due to the UK's size and economic power (particularly in the financial sector), it is likely that the UK will play an important role in developing trade policy, while relinquishing power in the fields of social and federal policy. One area of future research might be to model the level of engagement of a particular country, for example with the aid of a fuzzy variable.

References

Alonso-Meijide, J., Bowles, C., Holler, M., Napel, S.: Monotonicity of power in games with a priori unions. Theory Decis. **66**, 17–37 (2009)

Alonso-Meijide, J.M., Fiestras-Janeiro, M.G.: Modification of the Banzhaf value for games with a coalition structure. Ann. Oper. Res. **109**, 213–227 (2002)

Banzhaf, J.F.: Weighted voting doesn't work: a mathematical analysis. Rutgers Law Rev. **19**, 317–43 (1965)

Bertini, C., Freixas, J., Gambarelli, G., Stach, I.: Some open problems in simple games. Int. Game Theory Rev. **15**(02), 1340005 (2013)

Bertini, C., Gambarelli, G., Stach, I.: Apportionment strategies for the European parliament. In: Gambarelli, G., Holler, M.J. (eds.), Power Measures III. Homo Oeconomicus, vol. 22, pp. 589–604 (2005)

Gambarelli, G., Stach, I.: Power indices in politics: some results and open problems. Homo Oeconomicus **26**, 417–441 (2009)

Gambarelli, G., Stach, I.: Sub-coalitional approaches to values. In this edition (2017)

http://ec.europa.eu/eurostat/data/database. Accessed 21 Dec 2016

Johnston, R.J.: On the measurement of power: some reactions to Laver. Environ. Plann. A **10**, 907–914 (1978)

Kauppi, H., Widgrén, M.: Voting rules and budget allocation in an enlarged EU. Eur. J. Polit. Econ. **23**, 693–706 (2007)

Kóczy, L.Á.: Beyond Lisbon: demographic trends and voting power in the European Union Council of ministers. Math. Soc. Sci. **63**(2), 152–158 (2012)

Leech, D.: Computing power indices for large voting games. Manage. Sci. **49**, 831–837 (2003)

Malawski, M.: "Counting" power indices for games with a priori unions. Theory Decis. **56**(1–2), 125–140 (2004)

Mercik, J.W., Turnovec, F., Mazurkiewicz, M.: Does voting over national dimension provide more national influence in the European parliament than voting over ideological dimension? In: Owsiński, J.W. (ed.), Integration, Trade, Innovation and Finance: From Continental to Local Perspectives. Polish Operational and Systems Research Society, Warsaw, pp. 173–186 (2004)

Myerson, R.: Graphs and cooperation in games. Math. Oper. Res. **2**(3), 225–229 (1977)

Owen, G.: Values of games with a priori unions. In: Lecture Notes in Economic and Mathematical Systems, vol. 141, pp. 76–88 (1977)

Owen, G.: Modification of the Banzhaf-Coleman index for games with a priori unions. In: Holler, M.J. (ed.) Power, Voting, and Voting Power, pp. 232–238. Physica-Verlag, Würzburg (1982)

Owen, G.: Values of graph-restricted games. SIAM J. Algebraic Discrete Meth. **7**(2), 210–220 (1986)

Penrose, L.S.: The elementary statistics of majority voting. J. R. Stat. Soc. **109**, 53–57 (1946)

Roháč, D.: The unanimity rule and religious fractionalisation in the Polish-Lithuanian Republic. Const. Polit. Econ. **19**(2), 111–128 (2008)

Shapley, L.S., Shubik, M.: A method for evaluating the distributions of power in a committee system. Am. Polit. Sci. Rev. **48**, 787–92 (1954)

Turnovec, F.: Fairness and squareness: fair decision making rules in the EU council. Oper. Res. Decis. **4**, 109–124 (2009)

Van den Brink, R., Van der Laan, G.: A class of consistent share functions for games in coalition structure. Tinbergen Institute Discussion Paper No. 01–044/1 (2001)

www.warwick.ac.uk/~ecaae. Accessed 22 Dec 2016

Comparison of Voting Methods Used in Some Classical Music Competitions

Honorata Sosnowska[(✉)]

Warsaw School of Economics, Warsaw, Poland
honorata@sgh.waw.pl

Abstract. A comparison of the rules of voting in the last two main Polish classical music competitions: the XVIIth Chopin Piano Competition and the XVth Wieniawski Violin Competition. Weak and strong points of rules are analyzed. The rules are also compared to rules used in the previous editions of the competitions. We conclude that the changes resulted in the simplification of rules.

Keywords: Voting rules · Piano competitions · Violin Competitions · Outliers

1 Introduction

There are a lot of voting methods. They have different properties and are used in different situations. The outline of voting methods may be found in [1] or (in Polish) in [2]. Arrow's theorem and related topics show that there is no single method possessing "the best" properties [3]. Moreover, Gibbard and Satterthwaite's theorem [4, 5] shows that a strategical manipulability is almost always possible. There are special situations when experts assess competing projects or persons. In these situations usually some specific voting methods are used. Sometimes they are very complicated (see [6]), sometimes are constructed in a simple way. In some competitions a special computer program is used (the XIVth International Thaikovsky Competition, 2013 von Cliburn Piano Competition, 2013 Cleveland Piano Competition – Mc Bain's program). There are some papers about voting methods used in sport [2, 7–10] and in music [2, 6]. The aim of this paper is to analyze voting methods used in some classical music competitions and discover the direction of changes.

In this paper I compare voting methods used in the last two main Polish classical music competitions, the XVIIth International Fryderyk Chopin Piano Competition and the XVth International Henryk Wieniawski Violin Competition. I also compare rules of voting in these competitions with the rules used in previous editions. I decided to deal with the main Polish classical music competitions as this allowed a possibility to interview the organizers and read about the competitions in the local media. Rules of voting of Jury in XVIIth Chopin Piano Competition may be found on the Internet [11]. The rules of voting in the XIVth and XVth Wieniawski Violin Competitions are unpublished and were obtained for scientific purpose of writing this paper only.

The voting rules used in the XVIIth Chopin Competition and the XVth Wieniawski Competition are presented in Sects. 2 and 6 respectively. The social choice properties

© Springer International Publishing AG 2017
N.T. Nguyen et al. (Eds.): TCCI XXVII, LNCS 10480, pp. 108–117, 2017.
https://doi.org/10.1007/978-3-319-70647-4_8

of voting methods are analyzed in Sects. 3 and 7. The jurors' voting are analyzed in Sects. 4 and 8. Weak and strong points of these rules are discussed in Sects. 5 and 9. The rules are also compared to rules used in the previous competitions (Sects. 6 and 10). The paper ends with conclusions.

2 The XVIIth International Fryderyk Chopin Piano Competition

The XVIIth International Fryderyk Chopin Piano Competition took place between October 2–23, 2015. There were 78 competitors and 17 jurors. The competition consisted of 3 stages and a final. The rules stated that 40 competitors pass from stage I to II (in fact it was 43), 20 from stage II to III and no more than 10 to the final.

The following system of voting of jury was used. In stages I, II, III a double system was used. The system consists of Yes – No system and point system. The term "Yes – No" is used as a name for some modifications of the approval voting method. The approval voting method was introduced by Brams and Fisburn [12]. Using this method voters can choose as many alternatives as they want, 0, 1, 2, 3, even for all. In case of this competition jurors may assign "Yes" or "No" to as many candidates as they wish. The alternative approved by the greatest number of voters wins. In many competitions the method is called a Yes – No voting method. The Yes – No system is the main voting system in the XVIIth Chopin Piano Competition. The point system with scale 1–25 points (25 points – the best) plays an auxiliary role. It decides about the results should the Yes – No system fail. There is a possibility of discussion. Jurors know only anonymous results. Names of competitors may be known in stage III, but in that case there may only be a difference between the final list and the list constructed by the Yes – No voting on places 9–12.

When point system is applied, a problem of outliers arises. Some single very high or very low assessments may cause result expectably high or low. Some procedures of correction of outliers are used. In this competition outliers are reduced to the mean plus/minus a given deviation. There is computed an arithmetic mean m of all points. Results greater than $m + a$ or lower than $m - a$ are reduced to $m + a$ and $m - a$ respectively. The deviation is $a = 3$ in stage I and $a = 2$ in stages II and III. Then, the arithmetic mean of such corrected numbers of points is computed and it is the main result for a competitor. The competitor with the highest mean wins.

Jurors do not assess competitors who are their students (to be a "student" was defined in detail in the jury rules of voting). In case of Yes-No voting the percentage of "Yes" votes is computed and the competitor with the highest percentage wins. In case of the point system means are computed by dividing by the lowest number of jurors.

Only point system is used in the final. The scale is 1–10 (10 points – the best) and jurors prepare a ranking (weak order) of competitors distributing 55 points in such a way that only one competitor can get 10 points. Other number of points may be used many times. The correcting procedure used in the previous stages is also used with deviation a = 2. Jurors do not vote for their students. Then, a list is made based on the sum of points. The list is voted and must be accepted by majority 2/3 of voters.

If the list is not accepted, a very complicated final procedure is implemented. Let us see the respective part of the rules of jury [11].

"If the proposed verdict fails to receive such an approval, the Jurors will vote to award each main prize in writing, beginning with first prize. The order of voting will be determined by the aforesaid list, showing the averages of the scores. If a majority of the Jurors authorized to vote (those to whom the "S" relationship does not apply) votes in favor of awarding the prize concerned to the pianist according to the list, the prize will be awarded, and the Jury will vote to award the next prize. In the opposite case an analogous vote will be held to award this prize to the next pianist on the list. If a majority of the authorized Jurors votes "Yes", the prize will be awarded. In the opposite case, the Jurors will decide by an open vote whether to award this prize at all. If the Jurors decide not to award this particular prize, they will move on to a discussion and a vote on awarding.

If the procedure described above fails to identify the winner of a particular prize, and the Jury decides that this prize should be awarded, the Jurors will follow another procedure concerning the prize in question. Each Juror will write the name of his/her candidate for this prize on a card. In this case a Juror will be allowed to indicate his/her "student" but should also write the letter "S" and his/her vote will not be counted. The prize will be awarded to the candidate for whom more than half of the present, authorized Jurors (those to whom the "S" relationship does not apply) cast their votes. If the first vote fails to produce the winner of a prize, after the results are disclosed successive rounds of voting by ballot will be held for the persons indicated in the first round to eliminate the candidates who receive the lowest number of votes, until the desired result is achieved. In the event that only two candidates are left and neither receives the votes of a majority of the present, authorized Jurors, the Jury will decide by an open vote whether to award this prize to two pianists on an equal basis or not to award it at all."

The construction is very complicated and the procedure is difficult to use. As we show in Sect. 5, the construction may not work in case of some jurors' preferences.

3 Social Choice Properties of Voting in the Chopin Piano Competition

I shall analyze the point system and the final procedure. Yes – No system, as the approval voting method, has well known properties (see [13]) as unanimity, independence of irrelevant alternatives, strategic manipulability.

3.1 Point System

Let us study the impact of deviation a which is different in stage I than in the other stages and final. In Table 1 a case of two competitors (A and B) and three jurors (J1, J2, J3) is studied. I get different results for deviation $a = 2$ than for deviation $a = 3$. Assessments of jurors are presented in rows.

Let us also see that the correcting procedure may cause lack of monotonicity $(a = 2)$.

Table 1. Impact of size of a deviation

Juror	A	B	A'	B'	A''	B''
			$a = 3$		$a = 2$	
J1	19	20	19	20	19	20
J2	16	15	16	15.6	17	16.6
J3	22	21	22	21	21	20.6
Sum	57	56	57	56.6	57	57.2
Mean	19	18.6				
Result		A > B		$A' > B'$		$B'' > A''$

Let us analyze independence of irrelevant alternatives. Jurors assess competitors A and B, and C and D. Their assessments for A and B, and C and D are in the same order. Independence of irrelevant alternatives holds if social decision for A and B, and C and D are also in the same order for all individual preferences. In Table 2 there is presented an example of individual preferences which shows that the independence of irrelevant alternatives does not hold. Assessments of jurors are presented in rows.

Table 2. Independence of irrelevant alternatives

Juror (i)	A	B	$A', a = 2$	B'	$A'', a = 3$	B''
J1	21	18	21	18.3	21	18
J2	13	20	17	20	16	20
J3	23	23	21	22.3	22	23
Sum	57	61	59	60.6	59	61
Mean	19	20.3				
Result		B > A		$B' > A'$		$B' > A''$
Juror (ii)	C	D	$C', a = 2$	D'	$C'', a = 3$	D''
J1	21	18	21	18	21	18
J2	16	17	17.6	17	16.6	17
J3	22	22	21.6	21	22	22
Sum	59	57	60.2	56	59.6	57
Mean	19.6	19				
Result		C > D		$C' > D'$		$C'' > D''$

Now, let us study whether the unanimity is preserved by the group decision. I denote the finite set of jurors by J. I denote juror i's assessments of competitors A and B before correcting procedure by a_i, b_i, respectively. The assessments after the correcting procedure are denoted by a'_i, b'_i. Arithmetic mean of a_i is denoted by $m(A)$, of b_i by $m(B)$. The following lemma holds.

Lemma 1. If for all jurors i in J $a_i > b_i$ then for all deviations $a > 0$ $i a'_i > b'_i$.

Proof. $a_i > b_i$, so I have, $m(A) > m(B)$. If $m(B) + a < m(A) - a$ I get $b'_i \leq m(B) + a < m(A) - a \leq a'_i$ and $a'_i > b'_i$. Let us study a case $m(A) - a \leq m(B) + a$. Let us suppose that there exists I in J such that $b'_i \geq a'_i$. I consider the following cases for b'_i.

1. $b_i' < m(B) - a$, $b_i' > m(B) + a$, impossible by construction of a_i', b_i'.
2. $m(B) - a \le b_i' < m(A) - a$. Then $a_i' \le b_i' < m(A) - a$, impossible by construction of a_i'.
3. $m(A) - a \le b_i' \le m(B) + a$. Then $b_i' = b_i$, $a_i = a_i'$, impossible by $a_i > b_i$ and $b_i' \ge a_i'$.
4. $m(B) + a < b_i' \le m(A) + a$. Impossible by construction of b_i'. $\square$

It follows Lemma 1 that $m(A') > m(B')$ and unanimity is preserved.

3.2   Final Procedure

I consider 6 pianists and 18 jurors divided into 3 groups of 6 jurors with the same assessments. The final procedure may not lead to final results. The example is presented in Table 3 and is based on the Condorcet paradox. Assessments of groups of jurors are presented in columns.

Means of points given to first 9 pianists are the same and the final procedure does not work. If I change the number of jurors in group II from 6 to 5, the whole procedure is needed to obtain the final results and pianists A, B, C win. The time consistency of jurors' preferences is assumed.

Table 3. Failure of the final procedure

Number	Pianist	I group 6 jurors	II group 6 jurors	III group 6 jurors	Mean
1	A	9	3	5	5.66
2	B	9	3	5	5.66
3	C	9	3	5	5.66
4.	D	5	9	3	5.66
5.	E	5	9	3	5.66
6.	F	5	9	3	5.66
7.	G	3	5	9	5.66
8.	H	3	5	9	5.66
9.	I	3	5	9	5.66
10.	J	1	1	1	1

4   Jurors Voting in the Chopin Piano Competition

Jurors voting in the Chopin Piano Competition may be found on the Internet [14–17]. Analysis of voting shows that jurors preferred not strictly defined conditions of voting. They did use approval voting opportunities, as the number of votes "Yes" is often greater than the limit of persons in the next stage. Jurors wanted a connection between the point system and Yes-No system. They defined a border, but it was a fuzzy border of points and jurors' votes "Yes" or "No" in stages I, II, III (source [15]). For example juror Adam Harasiewicz in stage III voted "No" for Kozak with 20 points and "Yes" for Lu, also with 20 points. The limit of persons passing to the next stage was also fuzzy. It was stated in the rules that the number of competitors in stage II was to be no more than 40, while in reality there were 43. Jurors often used ties, even in the final where according to the rules a ranking of competitors was required. One of the jurors, Martha Argerich, awarded 2 competitors with 9 points, 1 with 6 points, 3 with 5 points and 4 competitors with 4 points. There was a possibility to vote strategically in the final, where jurors had 55 points to distribute among the competitors. So, they might increase the number of points for their best competitors in order to give them higher position. Such situation was not observed. Some of the jurors used less than 55 points. The scale of the point system was 1–25 in stages I, II, III and 1–10 in the final. Finalists collected high assessments in the previous stages, usually higher than 20. In the final the highest assessment was 10. It was a mental scale obstacle for jurors to award the best competitors with less points than they collected in preceding stages [18]. The problem is connected with some scale problems (see [19]).

5 Weak and Strong Points of Rules of Voting in the Chopin Piano Competition

The strong points of the rule of voting of Chopin Piano Competition seem to be these regulations which allow discussions and do not require strict solutions. Especially the possibility of discussion, approval voting method without additional restrictions and strict connections to the point system are the strong points of the voting rules. The final procedure, which is very complicated and for some jurors' preferences may not lead to final results is a weak point of the rule. The correcting procedure makes strategic voting less probable when the point system is used which is also an advantage of the rules. Although, strategic voting is possible in the final during the distribution of 55 points (not observed).

6 Comparison the Voting Rules of the XVIIth Chopin Piano Competition and the XVIth Chopin Piano Competition

The XVIth International Fryderyk Chopin Piano Competition took place in 2010. There were 13 jurors and 78 competitors. Properties of voting rules are described in [6]. The double system of voting was used: Yes – No system and point system. In fact, Yes-No System was strictly directed by the point system. There were no fuzzy borders between the number of points given by a juror and their "Yes" vote. A very complicated correcting procedure was used, based on reduction and elimination of outliers. In some cases such a procedure might lead to a situation where the voting system would not work. Moreover, for some preferences of jurors unanimity was not preserved. In the XVIIth Chopin Piano Competition connections between the Yes-No system and the point system were weaker. A quite different correcting procedure was introduced. The procedure was based on reducing outliers and did not lead to wrong social choice properties. More questions were subject to discussion.

7 The XVth International Henryk Wieniawski Violin Competition

The XVth International Henryk Wieniawski Violin Competition took place between October 8 and 23, 2016. There were 48 competitors and 12 jurors. In stage IV there were 13 jurors, however the chairman decided not to vote because one of the competitors was his student. The competition consisted of 4 stages. According to the rules no more than 24 competitors may pass to the IInd stage, 12 to the IIIrd, and 6 to the IVth. There was a possibility of increasing this number in case equal number of "Yes" votes in stages I and II was obtained by more competitors.

In stages I, II, III Yes – No system was used but contrary to the Chopin Piano Competition it was an approval voting method with some restrictions. Jurors voted "Yes" or "No" but the number of votes "Yes" had to be not greater than the number of competitors allowed to the next stage. Jurors who were teachers of competitors did not vote for them. Replacing a vote of a teacher Jury added an additional vote in accordance with majority of votes of the Jury. In case of a tie the Chairman's vote decided. Then the competitors were ordered from the highest to the lowest number of "Yes" votes. The competitor with the highest number of "Yes" votes was the best. Jurors also used a point system (scale 1–25). It played only an additional role, individual for each juror. Points were not aggregated into a jury decision, so no correcting procedure was needed. The results of use the point system were not archived, so there is no possibility of knowing them.

In stage IV, the final, a kind of Borda count was applied. The jurors ordered the competitors from the first place to the last one without ties. They did not assess by points but assigned places, so there was not a mental scale obstacle which occurred in the Chopin Piano Competition. The Chairman, vice-chairman and director of the competition assigned to each place a number of points equal to the number of place. Then the competitors were ordered with the increasing number of points. The competitor with the lowest number of points was the best. There was a difference in assigning places in reality and in the rules of competition. There rules stated that 6 points were to be assigned to the first place, 5 to the second, and finally 1 to the sixth [20]. It did not change properties of the voting system.

8 Social Choice Properties of the Wieniawski Violin Competition

The methods used by jury in the Wieniawski Violin Competition are well-known and very often used methods. Their properties are widely known (see e.g. [1, 4]).

9 Jurors Voting in the Wieniawski Violin Competition

Jurors voting in the XVth Wieniawski Violin Competition can be found on the Internet [21–24]. Jurors used some possibilities of Yes-No voting. Some of them did not use all their "Yes" votes (e.g. Vengerov, Bryla in the II stage). The possibility of increasing number of competitors passing to the next stage was used, even in the IIIrd stage, which was not expected in the rules. The voting system did not form outliers, so no correcting procedure was needed. In spite of this, there are some suspicions that jurors voted strategically in stage IV promoting one of the competitors ([25], in Polish). Cluster analysis provides some confirmation of this theory [26].

10 Weak and Strong Points of Wieniawski Violin Competition

A strong point of the rule of voting in XVth Wieniawski Violin Competition is the simple system of voting. Jurors did not have to know their full preferences in stages I, II, III, where they voted using the restricted Yes – No system. No problem of outliers appeared in any stage. Assignment of places, not points, in stage IV did not lead to the mental problem of scale (a more limited scale in the last stage than in the previous ones).

The upper limit for number of "Yes" votes in stages I, II, III can be viewed a weak point. Jurors had to calculate their votes, in case of stage I, 24 votes. Another weak point is the use of the Borda count in stage IV. It needs full (strong) order of competitors which may difficult for jurors (as was seen in the Chopin Piano Competition, some jurors like ties). Moreover, the Borda count may be prone to manipulation.

11 Comparison of Voting Rules of XIVth and XVthWieniawski Violin Competitions

The XIVth International Henryk Wieniawski Violin Competition took place in October 2011. There were 13–14 jurors and 53 competitors. The competition consisted of 3 stages and a final. In stages I, II and III each juror could award a competitor with at most 75 points, and up to 50 points in the final. The final result was a sum of points from all four stages. An elimination procedure was used. The highest assessment and the lowest assessment were eliminated. The assessments were not reduced. Table 4 demonstrates that this method may lead to different results than elimination. Jurors (J1, J2, J3) assessments are presented in columns. We see that violinist B (with significantly lower sum of points wins after the use the elimination procedure.

Table 4. Non-monotonicity of the voting method used in the XIVth Wieniawski Violin Competition

Violinist	J1	J2	J3	Sum	Sum after elimination the lowest and highest assessments
A	10	5	3	18	5
B	7	6	1	14	6

Lemma 2. Voting method used in the XIVth Wieniawski Violin Competition preserves unanimity.

Proof. Let violinist A get assessment $x_1,\ldots,x_k$ and violinist B $y_1,\ldots,y_k$ such that $x_i > y_i$ for all i. Let $x_1 = Min(x_1,\ldots,x_k)$, $x_k = Max(x_1,\ldots,x_k)$, $y_m = Min(y_1,\ldots,y_k)$, $y_n = Max(y_1,\ldots,y_k)$. We have y_1, $x_1 \leq x_m$, $y_k \leq y_n < x_n$ and after elimination of x_1, x_k, y_m, y_n the sum of other x's is greater than the sum of other y's.   □

An example situation for which independence of irrelevant alternatives does not hold is presented in Table 5. There are four jurors J1, J2, J3, J4. Their assessments are presented in columns. All jurors assess violinists A and B and C and D in the same order but the pianists are differently ordered by the group decisions.

Table 5. Irrelevance of independent alternatives in case of the voting method used in the XIVth Wieniawski Violin Competition does not holds

	J1, eliminated	J2	J3	J4, eliminated	Result
A	10	8	5	4	13
B	19	6	8	2	14
C	10	9	5	4	14
D	11	6	6	2	12

The above shows that the voting method used in the XIVth Wieniawski Violin Competition was more complicated than the one used in the XVth Wieniawski Violin Competition. The construction of both methods is quite different. The first is based on the point system, the second on the approval voting method and the Borda count. The method used in the XVth Wieniawski Violin Competition was chosen from many projects [20]. The organizers were looking to find "the best" system. As we know from Arrow's theorem and related topics this is impossible.

12 Conclusions

I analyzed the voting methods used in two main international classical music competitions which took place in Poland. In both cases the methods used in the last competitions were simpler than in the previous editions. Especially, less complicated methods of reducing impact of outliers are used (even a method which does not create outliers). Jurors used the advantages of the methods in their votes. When it was possible, they used ties. It seems that weak orders of competitors are more comfortable for jurors than strict orders. Simpler methods seem to be better, as they allow the jurors to anticipate results of their decisions.

References

1. Nurmi, H.: Comparing Voting Systems. D. Reidel Publishing Company, Dordrecht (1987)
2. Rzazewski, K., Slomczynski, W., Zyczkowski, K.: Kazdy glos się liczy (Each vote is counted, in Polish). Wydawnictwo Sejmowe, Warsaw (2014)

3. Arrow, K.J.: Social Choice and Individual Values, 2nd edn. Wiley, New York (1963)
4. Arrow, K.J., Sen, A.K., Suzumura, K.: Handbook of Social Choice and Welfare, vol. 1. Elsevier, Amsterdam (2002)
5. Gibbard, A.: Manipulation of voting schemes. Econometrica **41**, 587–601 (1973)
6. Sosnowska, H.: The rules for the Jury of the Fryderyk Chopin Piano Competition as a non standard voting rule. Roczniki Kolegium Analiz Ekonomicznych **32**, 23–31 (2013)
7. Gambarelli, G.: The "coherent majority average" for juries' evaluation processes. J. Sport Sci. **26**, 1091–1095 (2008)
8. Gambarelli, G., Iaquinta, G., Piazza, M.: Anti-collusion indices and averages for evaluation of performances and judges. J. Sports Sci. **30**, 411–417 (2012)
9. Tyszka, T., Wielochowski, M.: Must boxing verdicts be biased. J. Behav. Decis. Making **4**, 283–295 (1991)
10. Przybysz, D.: Agregowanie ocen sedziow sportowych jako przyklad zbiorowego podej-mowania decyzji (Aggregation of the (sport) judges marks as a form of collective decision making, in Polish). Studia Socjologiczne **1–2**, 105–136 (2000)
11. Web 1. http://static.eu.chopincompetition2015.com/download/regulamin_jury_konkursu.pdf. Accessed 21 Mar 2017
12. Brams, S., Fisburn, P.C.: Approval Voting. Birkhaser, Boston (1983)
13. Laslier, J.F., Sanver, M.R. (eds.): Handbook on Approval Voting. Springer, Heilderberg (2010)
14. Web 2. http://static.eu.chopincompetition2015.com/u299/i_etap_oceny.pdf. Accessed 21 Mar 2017
15. Web 3. http://static.eu.chopincompetition2015.com/u299/ii_etap_oceny.pdf. Accessed 21 Mar 2017
16. Web 4. http://static.eu.chopincompetition2015.com/u299/iii_etap_oceny.pdf. Accessed 21 Mar 2017
17. Web 5. http://static.eu.chopincompetition2015.com/u299/final_oceny.pdf. Accessed 21 Mar 2017
18. Oral communicate by the Chopin Competition organizers
19. Igersheim, H., Baujard, A., Gavrel, F., Lebon, L.I.: Individual behavior under evaluative voting: a comparison between laboratory and in situ experiments. In: Blais, A., Laslier, J.F., Van der Straeten, K. (eds.) Voting Experiments, pp. 257–269. Springer, Heildelberg (2016)
20. Oral communicate by the Wieniawski Competition organizers
21. http://www.wieniawski-competition.com/konkurs-skrzypcowy/wp-content/uploads/sites/3/2016/11/WYNIKI-I-ETAP-RESULTS-STAGE-1.pdf. Accessed 21 Mar 2017
22. http://www.wieniawski-competition.com/konkurs-skrzypcowy/wp-content/uploads/sites/3/2016/11/WYNIKI-II-ETAP-RESULTS-2ND-STAGE.pdf
23. http://www.wieniawski-competition.com/konkurs-skrzypcowy/wp-content/uploads/sites/3/2016/11/WYNIKI-III-ETAP-RESULTS-3RD-STAGE.pdf
24. http://www.wieniawski-competition.com/konkurs-skrzypcowy/wp-content/uploads/sites/3/2016/11/RANKING-FINALOWY-RANKING-LIST.pdf
25. http://wyborcza.pl/7,113768,20925157,konkurs-wieniawskiego-2016-wewnetrzna wo
26. Oral communicate by K. Kontek

Determinants of the Perception of Opportunity

Aleksandra Sus[(⊠)] [ID]

Wroclaw University of Economics, Komandorska Street,
53-345 Wroclaw, Poland
aleksandra.sus@ue.wroc.pl

Abstract. Contemporary strategic management has accepted the category of opportunity, although it cannot be reflected in the organization's plans and strategies. Alertness, proactivity, social networks and knowledge resources are the categories that come up most often when discussing opportunity perception as one of the determinants of entrepreneurial activity. In reality, they are the result of both behavioral and cognitive processes. The purpose of the article is to identify the primary factors that predetermine the idiosyncrasy of how opportunity is perceived by various persons, such as creativity, intuition, and divergent thinking. The article presents opportunity value chain, paraphrases the order of M.E. Porter's value chain and the A. Koestler's concept of 'bisociation'. The article also discusses the process of group decision making in terms of opportunity.

The article has been based on a study of the subject's literature, but the conclusions provide important directions that are utilitarian in nature.

Keywords: Opportunity · Chance · Opportunity perception · Intuition · Creativity · Groups in the organization

1 Introduction

From the viewpoint of strategic management, opportunity is a phenomenon that cannot be included in the strategic plan, as its occurrence cannot be planned by definition (although one can prepare oneself for such a situation, for example, by creating a redundancy of resources [13] or organizational slack), is unique, and its perception depends on the skills (of persons involved in an organization's activities in the fields of identifying, discovering, or noticing opportunities. So, the prerequisite for the process of opportunity perception is a thorough knowledge of a company's distinctiveness, regardless of the position occupied in it. There is no doubt about it that seeing the same phenomenon in different ways by various people is related to the nature of man and his individual characteristics. The idiosyncrasy of opportunity applies not only to the very phenomenon of opportunity, but also to its value. To some, certain situations will be of little value, to others, they may be a priority [22].

When reviewing the literature regarding entrepreneurship (both in general and in detail – strategic entrepreneurship), one may feel some cognitive dissonance caused by imbalance in the analysis of factors determining the process of opportunity perception (which, after all, precedes its utilization), which is most probably attributable to

© Springer International Publishing AG 2017
N.T. Nguyen et al. (Eds.): TCCI XXVII, LNCS 10480, pp. 118–128, 2017.
https://doi.org/10.1007/978-3-319-70647-4_9

orientation of its operationalization, leaving behavioral elements for sciences such as behavioral or social psychology.

It seems, however, that the cognitive aspects of the idiosyncrasy of opportunity should not be left without comment, especially as far as management science is concerned. So, how to analyze the factors determining the functioning of contemporary organizations (opportunities) without having detailed knowledge of how to solicit them? The purpose of this article is to give a better understanding of this subject, and the amorphousness of opportunity categories has provoked the structure of the publication.

The article consists of fourth parts. The first one provides an introduction into the problems of opportunity in strategic management. opportunity categories were conceptualized in an opportunity versus chance array, including threads of threats and situations indifferent to the organization. The second part of the paper aims at giving a better understanding of the auxiliary mechanisms underlying the process of opportunity management (key actions in the opportunity chain), which, in turn, result from behavioral factors, described in third part. The last section described determinants of group decision making in terms of opportunities, which is very substantial field but rarely communicated.

The considerations are theoretical, but the conclusions are utilitarian directions, which can have a significant effect on the functioning of contemporary organizations.

2 Opportunity Or Chance?

The encyclopedic description of the category of 'opportunity' defines it as a favorable system of circumstances leading to chances or possibilities, but also as a stroke of luck, which can be caught or lost [24]. Chance, in turn, has a broader meaning because [163]:

  i. it is a phenomenon that does not result from existing cause-and-effect relationships (sic!),
 ii. it is based on pointless, impersonal and unexplainable phenomena such as luck, coincidence, and destiny.

The German die Chance (chance) is used as the opposite of risk, whereas opportunity (German: die Gelengenheit) stresses the positive effects of a situation, without highlighting the related risk. Both meanings are rather positive in nature, unlike the English word 'chance', which can assume various forms, depending on the circumstances: there is a chance of achieving profit, but it can also involve loss [18].

R. Krupski has made an attempt to standardize terminology, giving 'chance' a high or low likelihood of success and 'opportunity' a tone of the occurrence of a certain event related to favorable conditions, rather without grading the potential advantage [14]. If a chance comes up, the category of threat has to be taken into account. And what if an opportunity turns out to be neither a chance nor a threat? Is it a neutral situation of no significance to the organization?

It is doubtful whether there can be any situations that occur in the environment of the organization which have no effect on the latter. According to social network analysis methodology, in addition to active factors, one can distinguish passive, critical, and idle, but never indifferent factors. Active factors have a very high effect on other elements, but are not subject to any influences; passive factors have little effect on others, but are subject to influences. Idle factors, in turn, have little effect on elements, but are also hardly subject to influences [28] (Fig. 1).

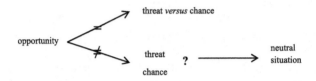

Fig. 1. Forms of opportunity Source: own work.

Following the track of the relationship 'entrepreneurship – opportunity', however, it should be made clear that these two elements are inseparable. For M. Bratnicki, chance is one of the eight key distinguishing features of entrepreneurship, which can be defined as discrepancy in an observed, constantly changing landscape and time image that has not been noticed and used by others. Such a gap can be filled by companies by providing their customers with certain values. No perfect chance exists because various parties can use it in different ways [2]. Subjectivity in differentiating between chances and threats is also emphasized by K. Obłój: '…a true chance and threat are like yin and yang, a ripple on water, or entanglement of good and evil. Evaluation of an ambiguous event or trend in terms of chance or threat is usually a matter of perspective, but not an objective assessment' [16]. This is also stressed by S. Shane and S. Venkataraman, according to whom opportunity perception is subjective, but opportunities themselves constitute an objective phenomenon [22].

The priority of opportunity perception is key to the success of entrepreneurs, which is stressed by I.M. Kirzner in his papers, and the introduction of disproportions in knowledge resources on the market is a major source of opportunities [9]. According to P.F. Drucker, one of the prerequisites for entrepreneurial management is the organization's ability to perceive changes in terms of opportunities, but not threats, which consequently changes the way of its functioning toward *rerum novarum cupidus* (that is, being eager for novelties) [4]. Opportunities reside both inside the organization and in its environment. According to I. Peiris, M. Akoorie and P. Sanh, opportunities are about existing and future customers, profits, and savings [17], while the internal forms of opportunity can be new products, services, means of production, and organizational methods, the existence of which brings higher profits than manufacturing costs [21]. And it is exactly the task of today's entrepreneurial strategic management to integrate dynamically the company's interior and environment, with the common denominator of such actions being opportunities.

The second section is discussed the opportunity value chain as a description of main and auxiliary actions which are critical in the process of opportunity perception.

3   Opportunity Value Chain

To paraphrase the order of M.E. Porter's value chain, auxiliary mechanisms that support the process of opportunity perception and main actions have been defined by studying the subject's literature (Fig. 2).

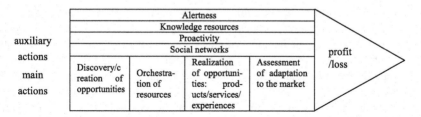

		Alertness		
auxiliary actions		Knowledge resources		
		Proactivity		
		Social networks		
main actions	Discovery/creation of opportunities	Orchestration of resources	Realization of opportunities: products/services/experiences	Assessment of adaptation to the market

profit /loss

Fig. 2. Opportunity value chain Source: own work.

The auxiliary actions have been recognized to include mechanisms that support the process of opportunity perception, that is, alertness, broadly understood knowledge resources, proactivity, and social networks. Alertness is the category introduced by I.M. Kirzner in 1973 and defined as openness to opportunities that are available but have been ignored so far. Opportunities are searched for without deliberately selecting techniques for scanning the environment. The prerequisite for being alert is a creative attitude to existing opportunity sources [7], whereas existing cognitive maps are the mechanism that inhibits opportunity perception.

Knowledge is utilized by the organizations in various forms and different ways. Pre-existing, historical knowledge of actions taken by the organization and the industry is significant to opportunity perception. The idiosyncrasy of pre-existing knowledge creates the so-called knowledge corridors, which enable an individual to identify specific opportunities [21]. According to S. Shane, there are three major dimensions of pre-existing knowledge which are key to the process of opportunity perception, and namely: pre-existing knowledge of markets, of the way in which services are provided on them, and of problems encountered by customers. In addition, the author stresses the need to draw attention, in the context of creating new products or services, to new information on technologies used [21]. The relationship between a higher level of pre-existing knowledge and the identification of a greater number of (more innovative) opportunities has been empirically confirmed by D.A. Shepherd and D.E. Detienne [23]. Knowledge also involves the use of one's own professional experience, which, the deeper it is, the more extensive final effects, including social networks, generates. When entering the issues of informal relations, on which social networks are also based, it is worth noticing that not so perception of opportunities as their utilization is by far easier when the decision maker has a well-established position in a relation. How easier can new orders can be solicited for the company when its owner is considered to be an honest and conscientious man who is readily recommended to other organizations? Building social networks furthers proactivity, that is, the ability to see cause-and-effect relationships between specific phenomena. It is important not only to respond to situations that are visible and understandable to the organization at a given time, but to anticipate changes that are consequences of such situations. In this setting, also the skills of noticing the weight of relational resources become significant.

Main actions, in turn, are stages of the process of perceiving opportunities that are discovered or created by the company. Four major phases of this process have been identified by studying the subject's literature, namely:

i. Discovery/creation of opportunities; as mentioned above, it is the stage around which a heated discussion is held on the subject's literature; it seems, however, that, regardless of whether opportunities are identified or created by the entrepreneur, the most important is the value related to their utilization and the entrepreneur's individual predispositions, which predetermine the effectiveness of this stage;

ii. Orchestration of resources, that is, efficient management of resources, which involves the processes of acquiring and selling them, maintaining certain redundancies and an appropriate structure, and being alert to capture moments in which such slack has to be released or withdrawn; W. Czakon names orchestration of resources as one of the five elements of relational skills, that is, those that are a strategic distinguisher of cooperating businesses, which, due to the patterns they follow through common utilization of resources, achieve better results than competitors [3].

iii. Realization of opportunities, which can assume the form of new products, services, or experiences,

iv. Assessment of adaptation to the market.

The concept of M.E. Porter's value chain is a certain simplification of the opportunity management procedure. They were rather used to visualize those mechanisms that are auxiliary, but are necessary for opportunity perception and the main stages of the opportunity utilization process (here, in turn, with an inclination toward the classic management functions). The analogy and effects of the discussed process in the form of new products, services or experiences have also enabled indication of an element that is omitted in the literature on entrepreneurship, and namely opportunity marketing.

Nevertheless, the considerations to date do not provide an answer to what causes that only some entrepreneurs use the auxiliary mechanisms for opportunity perception? Extremely important for the answer that question is to focused on heuristics that entrepreneurs use – this is the goal of the third section of the article.

4 Heuristic Determinants of Opportunity Perception

G.E. Hills's and G.T. Lumpkin's research shows that creativity is key to the process of opportunity perception, both declared by respondents (more than 50% of entrepreneurs and their representatives. The test sample consisted of 53 entrepreneurs and their 165 representatives in organizations with an annual income of more than 3 million U.S. dollars and an employment of 2 to 1,100 persons) and in their subjective self-evaluation. Creativity is the feature that has predetermined their successes, and intuition and life optimism its main elements [6]. But at which stage of opportunity management? Of their perception or utilization, or perhaps of auxiliary mechanisms? According to the model shown in Fig. 3, entrepreneurial creativity is the result of associative and bisociative thinking within recognized social networks, alertness, pre-existing experience, and experience.

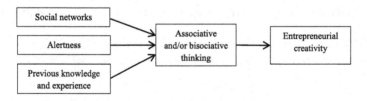

Fig. 3. Model of entrepreneurial creativity Source: based on: [11]

The theories of association are based on thinking through analogy, creation of effects through associations of two pieces of information that exist independent of each other. A. Koestler has introduced the concept of 'bisociation' to make a clear distinction between routine thinking skills M1 (perfectly logical) and act of creativity M2 (unexpected and surprising) [11] (Fig. 4).

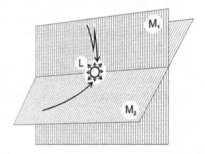

Fig. 4. Matrices of linear and lateral thinking in the bisociation model Source: [11]

The occurrence of situation L (also analyzed opportunities) results in a simultaneous vibration and intermingling of two different wavelengths – hence the name of the theory – causing instability between feelings (emotions) and reason (prudence). When this mechanism is translated into opportunity management processes, a conclusion can be made about the relationships between the idiosyncrasy of opportunities and the wavelength of non-linear thinking. The longer this wavelength is, the later the aspects of logical thinking, which, by definition, eliminate elements of creativity such as instinct, emotions, and fantasy, become important.

In 2002 D. Kahneman and A. Tversky won the Nobel Prize for economics. His works assume that human decisions are not based on logic, but irrational actions (Kahneman 2002, Prize Lecture, prepared jointly with A. Tversky). The results of heuristic research also indicate that the greatest business successes are attained by entrepreneurs who feature high innovativeness in action and tend to use heuristics in their decision-making processes. These heuristics provide important clues about how to act in times when analyses lose their validity already when they are being done [25] and the author's own observations):

1. Intuitive decision making; following one's own opinion and rejecting expert judgments are immanent features of entrepreneurs who enjoy spectacular successes.
2. Focus on the company's growth and elimination of those actions and decisions that do not lead to it.
3. Emotional attachment to the organization. The company is not just work, it is the entrepreneur's 'second home' to which the entrepreneur dedicates their time and energy.

Orientation to profit only is also disputable. As already mentioned above, the entrepreneur is guided rather by the will to grow, and only afterward by profit (which is not always a natural consequence of the company's growth).

4. Orientation to people, who are not only employees for the entrepreneur. They are also fellow decision makers and partners. Entrepreneurs are not opportunists.
5. Acting according to one's own vision. And here it should be agreed with Professor R. Krupski's opinion that opportunity is filtered by nothing but vision.

Based on analysis of the subject's literature, available research results and the author's own observations, it can be ventured to claim that in the process of opportunity perception entrepreneurs found their actions on heuristic processes. This means that mechanisms related to rather subconscious than strictly rational functioning should underlie all opportunity management processes. These mechanisms also predetermine the idiosyncrasy of both opportunities and their value. Only afterward do processes based on cognitive mechanisms take place, which trigger existing patterns that enable entrepreneurs to join the dots. Or they occur simultaneously, in accordance with the theory of bisociation.

The next very interesting feature in the area of the perception of opportunity is fact that its determinants in the literature are always analyzed via organizations, but not viewed through the individuals or groups. In the fourth section of the article attention was focused on the determinants of group decision making in terms of opportunities.

5 Determinants of Group Decision Making in Terms of Opportunities

E. Schein defines a group as any number of people who: (a) are connected by interactions, (b) psychologically self-aware (c) they perceive themselves as a group and (d) have a common goal, expressed in the type of activity and expected performance standards thereof. Achieving the goal can involve solving an organizational problem, taking a business decision, as well as sharing information and formulating new ideas [12]. In addition, the existence of the group is contingent on creating a structure consisting of group positions, the hierarchy of power, a sociometric system and communication ties between the members of the group [15].

Analyzing the differences in the identification and exploitation of opportunities by the members of the group should particularly involve the perception of potential risks. Membership in the group has substantial advantages, which also pose a threat to the effective use of opportunities. The group minimizes the dangers of loneliness, people become stronger and more confident, and their self-esteem grows. The need of belonging is satisfied, and things that used to be impossible or difficult to achieve individually, become possible through interactions between members of the group [20]. This can radically change the perception of opportunity, and improve the chances of new, emerging situations being perceived as such. This is relatively risky, because the organizational culture becomes focused on making entrepreneurial decisions. Consequently there is competition between the members of the group, which can potentially lead to erroneous assessment of opportunities. In such situations the leader of the group plays an important role and ensures the right balance between the benefits and negative consequences of group functioning in the opportunity focused environment.

The category which determines the group decision-making in terms of opportunity is the composition of the group, that is, the mutual connections of its members and the level of motivation to stay in the group. The more close-knit the group is, the more effectively its members implement the objectives and achieve greater results in the identification and exploitation of opportunities.

This is possible by: (a) reducing the size of the group, (b) increasing the time spent together (without interfering with the free time, due to possible conflicts), (c) encouraging competition with other groups, or (d) rewarding the whole group rather than its individual members [20].

In terms of opportunity the size of the group is an important determinant of its effectiveness. This, in turn, depends on the tasks that have been put before the group to carry out, and variables such as personality traits of group members and characteristics of its leader.

When fast reaction is needed, a small group (3–4 members) may be fully effective. Such groups are easier to control, and are characterized by greater coherence. There is also a smaller risk of conflicts. The drawback of such a small group could be the lack of specific skills and competences [12].

The survival of the group depends on the creation of relationships between its members. In the case of triads, or groups of three, there are several possible relationships between their members, such as coalitions of two people against one, or conflicts. In this case, the third member of the group can play the role of a mediator, whose task is to avert conflict. A triad is more stable than a couple, but less stable than a large group. With the increase in size of the group ties between its members become blurred [15].

Social idleness may be a significant threat to large groups, especially in the case of teams focused of opportunity. In such a situation a reduction of the size of the group should be considered to make it possible to measure the amount of work that individual members contribute [20].

To sum up, there are certain differentiators that determine the effectiveness of the group in the identification and exploitation of opportunities, but a properly constructed collective decision-making unit allows to achieve many benefits for the organization. As the saying goes "two heads are better than one", the group allows for creation of many different possible solutions to the problem, and consequently there is a much greater probability of identifying unforeseen developments in the environment of the organization (opportunities). Teamwork releases energy and fosters creativity, which in turn allows to use a wider range of knowledge and experience. However, groups can also favor hasty solutions, and rash decisions resulting from various arrangements/agreements between the group members. In addition, a dominating member can undermine group effort (false unanimity), while strong members can influence others and lead to wrong solution to the problem. In the group there is a high risk of occurrence of conflicts that can not only impede performance of the group but also, in the long term, influence the ability to perceive opportunity. Too long time that members spend on communication (too much talk, too little action) also reduces the quality of group decision.

In the case of decisions taken in groups, the presence and influence of other people can radically influence the process of identification and taking advantage of an opportunity. Definitely less significance can be attributed to heuristics, and rational decision-making characteristics prevail.

6 Summary

As previously mentioned, contemporary strategic management has accepted the category of opportunity. Increasingly more companies even retain certain margins of various resources, which they can use in case of opportunity. The question should be asked, however: why do not organizations have so advanced 'opportunity' procedures as they do in case of crises? Is this caused by the pejorative nature of crises? What makes organizations not 100% prepared for an opportunity to come up? It is disputable, after all, whether it is possible at all to get prepared for opportunities? Especially in the context of their irrational foundations. In the said work *The Art Of Creation*, A. Koestler writes: '*Everyone can ride a bicycle, but no one knows how to do it*' [11]. It is similar with opportunities. Providing a better understanding of mechanisms responsible for their utilization and idiosyncratic perception is, in fact, only a mere description of a certain existing condition. It is doubtful, however, whether it is possible to learn how to become an entrepreneur. In particular, considering the complexity levels of factors that predetermine their idiosyncrasy and, above all, their original, that is, unconscious character (as seen by Z. Freud).

References

1. Baron, R.A.: Opportunity recognition as pattern recognition. how entrepreneurs 'Connect the Dots' to identify new business opportunities. Acad. Manag. Perspect. **20**(1), 108 (2006)
2. Bratnicki, M.: Przedsiębiorczość strategiczna w strategicznej mozaice czasu [Strategic Entrepreneurship in Strategic Time Mosaic]. In: Grudzewski, W.M., Hejduk, K.I. (eds.) Przedsiębiorstwo przyszłości – wizja strategiczna [Future Enterprise – Strategic Vision], Difin, Warsaw, pp. 129–130 (2002)
3. Czakon, W.: Sieci w zarządzaniu strategicznym [Networks in Strategic Management], pp. 159–162. Oficyna a Wolters Kluwers Business, Warsaw (2012)

4. Drucker, P.F.: Innowacje i przedsiębiorczość [Innovation and Entrepreneurship], pp. 163–168. PWE, Warsaw (1992)
5. Kahneman, D., Tversky, A.: A Maps of Bounded Rationality: A Perspective on Intuitive Judgment and Choice. Prize Lecture, pp. 449–489 (2008). http://www.nobelprize.org/nobel_prizes/economic-sciences/laureates/2002/kahneman-lecture.html
6. Hills, G.E., Lumpkin, G.T.: Opportunity recognition research: implications for entrepreneurship education. Paper Presented at İntEnt 1997, The 1997 International Entrepreneurship Conference, Monterey Bay, California, U.S.A., pp. 7–11 (1997). ftp://ns1.ystp.ac.ir/ystp/1/1/root/data/pdf/entrepreneurship/hills.pdf
7. Kirzner, I.M.: Competition and Entrepreneurship, p. 151. University of Chicago Press, Chicago (1973)
8. Kirzner, I.M.: Entrepreneurial discovery and the competitive market process: an Austrian approach. J. Econ. Lit. 35(1), 71–72 (1997)
9. Kirzner, I.M.: Konkurencja i przedsiębiorczość [Competition and Entrepreneurship], pp. 69–84. Fijorr Publishing Company, Warsaw (2010)
10. Ko, S.J., Butler, E.: Creativity: a key link to entrepreneurial behavior. Bus. Horiz. 50(5), 365–372 (2007)
11. Koestler, A.: Act of Creation, pp. 35–42. Hutchinson & Co. Ltd., London (1964)
12. Kożusznik, B.: Kierowanie zespołem pracowniczym [Managing the Team], pp. 15–20. PWE, Warsaw (2005)
13. Krupski, R.: Przedsiębiorstwo w ruchu [The enterprise in move]. In: Skalik, J. (ed.) Prace Naukowe AE we Wrocławiu nr 1045, [Research Papers of the WUE no. 1045], Wydawnictwo AE im. Oskara Langego we Wrocławiu, Wroclaw, pp. 82–83 (2004)
14. Krupski, R.: Rodzaje okazji w teorii i praktyce zarządzania [Types of opportunities in the theory and practice of management], Prace Naukowe Wałbrzyskiej Wyższej Szkoły Zarządzania i Przedsiębiorczości, [Research Papers of the WCME], vol. 21, Wałbrzych, p. 6 (2013)
15. Moczydłowska, J.: Zachowania organizacyjne w nowoczesnym przedsiębiorstwie, [Organizational Behavior in a Modern Enterprise], pp. 71–79. Wydawnictwo Śląsk, Katowice (2006)
16. Obłój, K.: Strategia organizacji [Organisational Strategy], p. 227. PWE, Warszawa (2007)
17. Peiris, I., Akoorie, M., Sanha, P.: Conceptualizing the process of opportunity identification in international entrepreneurship research. South Asian J. Manag. 20(3), 10 (2013)
18. Purvis, K.: Risk, Chance And Opportunity, Versicherungswirtschaft 15, p. 31 (2013). http://web.ebscohost.com/ehost/detail?vid=4&sid=df99d619-fe48-4a64-b7d1-ab75b21192be%40sessionmgr114&hid=108&bdata=Jmxhbmc9cGwmc2l0ZT1laG9zdC1saXZl#db=bth&AN=89593175
19. Renko, M., Shrader, R.D., Simon, M.: Perception of entrepreneurial opportunity: a general framework. Manag. Decis. 50(7), 1233–1251 (2012)
20. Robbins, S.P.: Zachowania w organizacji [Behavior in Organizations], pp. 169–178. PWE, Warsaw (2004)
21. Shane, S.: Prior knowledge and the discovery of entrepreneurial opportunities. Organ. Sci. 11(4), 451–452 (2000)
22. Shane, S., Venkataraman, S.: The promise of entrepreneurship as a field of research. Acad. Manag. Rev. 25(1), 220–221 (2000)
23. Shepherd, D.A., Detienne, D.R.: Prior knowledge, potential financial reward, and opportunity identification. Entrepreneurship Theor. Pract. 29, 91–112 (2005)
24. The New Lexicon Webster's Encyclopedic Dictionary of the English Language, Deluxe edn., p. 163, 704. Lexicon Publications, Inc., New York (1991)

25. Tyszka, T.: Psychologia ekonomiczna [Economic Psychology], pp. 319–320. Gdańskie Wydawnictwo Psychologiczne, Gdańsk (2004)
26. Venkataraman, S.: The distinctive domain of entrepreneurship research: an Editor's perspective. In: Katz, J., Brockhaus, R. (eds.) Advances in Entrepreneurship, Firm Emergence, and Growth, pp. 228–247. JAI Press, Greenwich (1997)
27. Wan, W.W., Yiu, D.W.: From crisis to opportunity: environmental Jolt, corporate acquisitions, and firm performance. Strat. Manag. J. **30**, 794 (2009)
28. Zimniewicz, K.: Współczesne koncepcje i metody zarządzania [Contemporary concepts and methods of management], p. 143. PWE, Warsaw (2009)

Free-Riding in Common Facility Sharing

Federica Briata[1] and Vito Fragnelli[2(✉)]

[1] Department of Mechanical, Energy, Management and Transportation Engineering,
University of Genova, Genova, Italy
federica.briata@libero.it
[2] Department of Science and Technological Innovation,
University of Eastern Piedmont, Vercelli, Italy
vito.fragnelli@uniupo.it

Abstract. We deal with the free-riding situations that may arise from sharing maintenance cost of a facility among its potential users. The non-users may ask for a check to assess who the users are, but they have to pay the related cost; consequently, a non-user may not ask for the check, with the hope that the other non-users ask and pay for it. In this paper, we provide incentives for asking for the check, without suffering a higher cost

Keywords: Free-riding · Mechanism · Fairness · Quorum

1 Introduction

In Briata [1] a non-cooperative game theoretical approach is introduced for managing the sharing cost of a facility available to several potential users; if someone actually does not use it, the trivial equal sharing of the maintenance cost of the facility is unfair because the non-users are charged for a service they do not exploit; in order to increase the fairness, the non-users have the possibility of asking that a check for establishing who the users are is made, but they have to pay its cost; the solutions of the game may be unfair and cause free-riding situations that may arise when non-users pay for the cost of users and when a non-user decides not to ask for the check, with the hope that the other non-users ask and pay for it. In this paper, we deal with these free-riding situations providing incentives for asking for the check, without increasing the cost. In particular, the situation analyzed in [1] is how to share the maintenance cost of a printer available to all the members of a Department. The easiest solution to the problem of cost division is equally sharing it among all the members, especially when no further information is available, but the users behave as free-riders w.r.t. the non-users. It is fairer to consider the actual users and the intensity of their use. It is common knowledge who the users are, but it is necessary an external intervention (supervision or some device) to verify who the users are and an additional cost has to be paid. This may be viewed as an information cost. Briata disregards the intensity of use of the facility, assuming that the maintenance

© Springer International Publishing AG 2017
N.T. Nguyen et al. (Eds.): TCCI XXVII, LNCS 10480, pp. 129–138, 2017.
https://doi.org/10.1007/978-3-319-70647-4_10

cost and the check cost are fixed and known to all the agents; moreover, they are comparable, e.g. they refer to the same period of time. The cost for getting the information was already considered in Moretti and Patrone [4] in a cooperative situation, leading to TU games with information cost, or TUIC games, that is a family of cost games with an additional cost to get the information about each coalition. Briata introduces the *naming* procedure: each player may *name* the facility, that is may ask for the assessment of who are the players that actually use the printer. In order to make or not the check, she assumes that the facility has to be named by at least one player, leading to *Naming Games*, or by a majority of the players, leading to *Majority Decision Games*. In a more general setting, we may suppose that the check is made if and only if the number of check requests is larger than a given threshold, and, in this case, the cost is equally paid by the players who asked for it, while the costs for the use of the facility is divided among the users only; otherwise, no check is made and all the players equally share the maintenance cost. As we already said, this situation is exposed to free-riding behavior of the users, if the check is not made, or of the non-users, if the check is made.

In a non-cooperative setting the main aim of each player is to minimize the amount s/he has to pay at the expense of fairness; so, for each user not to ask for the check is the best choice whatever the other agents do (dominant strategy); in fact, if the check is made, there is a reduction of the number of players among which the maintenance cost is divided, moreover the agents that asked for the check have to pay a quota of the related cost. The situation is more complicated for a non-user; in this case, the amount that has to be paid, depends not only on the choices of the other agents, that is normal in a game theoretic approach, but also on the maintenance cost, the check cost and the number of users and non-users; note that these data may be known or not. At a first glance, it may seem that the cost incurred by a non-user is the equal share of the maintenance cost among all the agents when the check is not done, while it is the equal share of the check cost among all the non-users when the check is done. This is not necessarily true. In fact, even when the share of the check cost is lower than the share of the maintenance cost, a non-user may not ask for the check in order to save money, hoping that the request and the payment of the check is up to other agents, incurring the risk that the check is not done, in other words we face a second free-riding problem. The non-users have a larger incentive for the free-riding behavior, the smaller is the number of agents required to obtain the check and the larger is its cost.

In this paper, we have two aims and we analyze both the situations in a non-cooperative setting. First, we look for reducing the advantage of a free-riding behavior of both the users and the non-users, modifying the rules for making the check and the rules for sharing the maintenance cost and the check cost; second, we want to improve the fairness of the cost allocation problem, looking for a game that, on the one hand, provides incentives to check also the intensity of use of the facility, and on the other hand, assigns to each agent a quota of the total cost that is satisfactory for as many agents as possible.

The paper is organized as follows. In Sect. 2 we provide the basic notion and notations of Game Theory; in Sect. 3 the models introduced in [1] are shortly outlined, referring to a new example; Sect. 4 re-analyzes the model in [1] in a non-cooperative setting defining two mechanisms that reduce the free-riding behavior of the agents; more precisely, in Subsect. 4.1, we refer to a check that allows only to distinguish among users and non-users, then in Subsect. 4.2, we suppose that the check may provide information about the intensity of usage of the facility; Sect. 5 concludes.

2   Basic Definitions and Notations

In this section, we introduce some standard terminology and some notations, for convenience. Since we are facing a cost allocation problem, we represent the situation by a cost game. Let $N = \{1, 2, ..., n\}$ be the finite set of players; we denote the cost game in strategic form by $(X_1, X_2, \ldots, X_n, c_1, c_2, \ldots, c_n)$ where X_i is the non-empty set of strategies of player $i \in N$ and $c_i : \prod_{k \in N} X_k \longrightarrow \mathbb{R}$ is the cost function of player $i \in N$, i.e. $c_i(x_1, x_2, \ldots, x_n)$ represents the amount that player $i \in N$ has to pay when the strategy profile $(x_1, x_2, \ldots, x_n)$ is chosen.

We denote by X_{-i} the set $\prod_{k \neq i} X_k$, by x_{-i} an element of X_{-i}, and by (y_i, x_{-i}) the element of $\prod_{i \in N} X_i$ obtained from $(x_1, x_2, \ldots, x_n)$ replacing the strategy x_i of player i with y_i.

Definition 1. *Given a cost game in strategic form* $(X_1, X_2, \ldots, X_n,$ $c_1, c_2, \ldots, c_n)$, *the strategy* $x_i \in X_i$ *strongly dominates the strategy* $y_i \in X_i$ *for player* $i \in N$, *if for each* $x_{-i} \in X_{-i}$

$$c_i(x_i, x_{-i}) < c_i(y_i, x_{-i}).$$

The strategy $x_i \in X_i$ *weakly dominates the strategy* $y_i \in X_i$ *for player* i, *if for each* $x_{-i} \in X_{-i}$

$$c_i(x_i, x_{-i}) \leq c_i(y_i, x_{-i})$$

and there exists a strategy $\bar{x}_{-i} \in X_{-i}$ *such that*

$$c_i(x_i, \bar{x}_{-i}) < c_i(y_i, \bar{x}_{-i}).$$

The strategy $x_i \in X_i$ *is strongly dominant for player* i, *if* x_i *strongly dominates every strategy* $y_i \in X_i$ *with* $x_i \neq y_i$, *while the strategy* $x_i \in X_i$ *is strongly dominated if there exists a strategy* y_i *which strongly dominates it.*

When the players cannot subscribe binding agreements we have a non-cooperative game. The most important solution for a non-cooperative game is the Nash equilibrium [6], that for a cost game is defined as follows.

Definition 2. *Given a cost game in strategic form* $(X_1, X_2, \ldots, X_n, c_1,$ $c_2, \ldots, c_n)$, *the strategy profile* (x_i^*, x_{-i}^*) *is a Nash equilibrium iff*

$$c_i(x_i^*, x_{-i}^*) \leq c_i(x_i, x_{-i}^*), \quad \forall \, x_i \in X_i, \forall \, i \in N$$

In other words, a strategy profile is a Nash equilibrium if and only if no player has an advantage in unilaterally deviating from it.

We consider a finite set of players $N = \{1, 2, \ldots, n\}$, which can use a facility. Let $M \subseteq N$ be the set of the players *not* using the facility. Assuming that $M \neq N$ is non-restrictive. Let $C > 0$ be the maintenance cost that we assume given, let $\chi > 0$ be the check cost, that is the cost to make the set of users of the facility verifiable, and let t be the threshold for obtaining the check. Each agent has two strategies, $\mathcal{A}$ and $\mathcal{N}$, where $\mathcal{A}$ stands for *asking for check* and $\mathcal{N}$ for *not asking for check*. For a user, the strategy $\mathcal{A}$ is strongly dominated by the strategy $\mathcal{N}$, as s/he has to pay a larger quota of the printer cost, plus a quota of the check. Removing strictly dominated strategies, i.e. all the agents in the set of users $N \setminus M$ choose the strategy $\mathcal{N}$, the remaining players in the set M are symmetric and the resulting game is a binary one, so it has at least one Nash equilibrium (for the proof, see [1]).

3   Recalls

In this section we recall via an example the main features of the games presented in Briata [1] and defined in a non-cooperative setting, addressing to the paper for further details.

Example 1. *Let* $N = \{1, 2, 3, 4, 5\}, M = \{1, 2, 3\}$ *and* $t = 2$. *As the users choose the dominant strategy* $\mathcal{N}$, *and the non-users are symmetric, there are four meaningful situations:*

a. *no agent asks for the check;*
b. *one non-user asks for the check and it is not made;*
c. *two non-users ask for the check and obtain it;*
d. *three non-users ask for the check and obtain it.*

The choices of players in N *and their costs are summarized in the following table, for the different cases.*

case	choice		individual cost	
	non-users	users	non-users	users
a	$\mathcal{N}\ \mathcal{N}\ \mathcal{N}$	$\mathcal{N}\ \mathcal{N}$	$\frac{C}{5}\ \frac{C}{5}\ \frac{C}{5}$	$\frac{C}{5}\ \frac{C}{5}$
b	$\mathcal{A}\ \mathcal{N}\ \mathcal{N}$	$\mathcal{N}\ \mathcal{N}$	$\frac{C}{5}\ \frac{C}{5}\ \frac{C}{5}$	$\frac{C}{5}\ \frac{C}{5}$
c	$\mathcal{A}\ \mathcal{A}\ \mathcal{N}$	$\mathcal{N}\ \mathcal{N}$	$\frac{\chi}{2}\ \frac{\chi}{2}\ 0$	$\frac{C}{2}\ \frac{C}{2}$
d	$\mathcal{A}\ \mathcal{A}\ \mathcal{A}$	$\mathcal{N}\ \mathcal{N}$	$\frac{\chi}{3}\ \frac{\chi}{3}\ \frac{\chi}{3}$	$\frac{C}{2}\ \frac{C}{2}$

In cases a. and b. all the agents pay $\dfrac{C}{5}$ *each; in case c. the two non-users that ask for the check pay* $\dfrac{\chi}{2}$ *each, the third non-user pays nothing and the two users pay* $\dfrac{C}{2}$ *each; in case c. the three non-users pay* $\dfrac{\chi}{3}$ *each and the two users pay* $\dfrac{C}{2}$ *each.*

We may remark that case a. always corresponds to a Nash equilibrium, because if only one agent switches from $\mathcal{N}$ to $\mathcal{A}$ the situation is the same as $t = 2$, and that case d. never corresponds to a Nash equilibrium because each non-user may improve her/his payoff not asking for the check and paying nothing profiting that the two other non-users asked for the check; case b. corresponds to a Nash equilibrium when $\frac{\chi}{2} \geq \frac{C}{5}$, and case c. corresponds to a Nash equilibrium when $\frac{\chi}{2} \leq \frac{C}{5}$ (note that for $\frac{\chi}{2} = \frac{C}{5}$ both cases b. and c. lead to a Nash equilibrium). Summarizing, we have:

1. $\dfrac{C}{5} \geq \dfrac{\chi}{2}$

   The non-users may obtain an advantage if the check is made; if only one non-user does not ask for it, s/he pays nothing (free-riding);

2. $\dfrac{\chi}{2} > \dfrac{C}{5} > \dfrac{\chi}{3}$

   The non-users obtain an advantage when the check is made only when all of them ask for the check; if only one non-user does not ask for it, s/he pays nothing (free-riding) and the other two pay more (inefficiency);

3. $\dfrac{\chi}{3} \geq \dfrac{C}{5}$

   Not to ask for the check is a weakly dominant strategy for the non-users.

If the threshold were $t = 1$, the number of Nash equilibria decreases, since case a. is no longer a Nash equilibrium if $\chi < \frac{C}{5}$.

The previous results may be generalized for all N, M, t, C and χ.

In the following section, we propose a mechanism that allows avoiding both free-riding and inefficiency.

4   The Non-cooperative Approach

In this section, we reconsider the model in [1] with a twofold objective: reducing the free-riding behavior and increasing the profitability of the check. In Subsect. 4.1, we introduce two mechanisms in order to make less profitable the free-riding behavior; in Subsect. 4.2, we extend the situations in which the agents prefer asking for the check, adding the condition that the result is not only a list of who the users are, but also how much they used the facility. The idea of mechanism design, known also as *reverse Game Theory* because it provides incentives in order to make easier to reach desired objectives, was developed, among the others, by Vickrey [8], Hurwicz [2], Myerson [5], Roth [7] Maskin [3], all awarded with the Sveriges Riksbank Prize in Economic Sciences in Memory of Alfred Nobel.

4.1   The Quorum

From the previous section, it is clear that the check quota to be paid by a non-user can be greater than the quota of the maintenance cost as a consequence of free-riding. In this case the check very probably is not made (see case $c.$ in Example 1 when $\frac{\chi}{2} > \frac{C}{5}$). In fact, strategy $\mathcal{A}$ is convenient in order to avoid the free-riding behavior of the users, but check cost can be too expensive if the number of users who ask for it is too scant. The idea of choosing an adequate threshold for making the check guarantees that it is made only if it is non-disadvantageous.

For this reason, we introduce the *quorum* q, i.e. the minimum number of agents that has to ask for the check accounting the cost C, the cost χ and the total number of agents n:

$$q = \min_{p \in \mathbb{N}_>} \left\{ p : \frac{\chi}{p} \le \frac{C}{n} \right\}.$$

Mechanism 1. *Fix the threshold equal to the quorum.*

Using Mechanism 1, if the quorum is reached, each agent who asked for the check pays no more than the equal share of the maintenance cost C, otherwise s/he pays the equal share of C. For instance, in Example 1, if $\frac{\chi}{2} > \frac{C}{5}$ then $q > 2$. But there is still the problem that the non-users may have a free-riding behavior, as the following example shows.

Example 2. *Let $N = \{1, 2, 3, 4\}$, $M = \{1, 2\}$, $C = 12$, $\chi = 2$; in this case $q = 1$, i.e. one agent is sufficient for obtaining the check. Suppose that the agents in the set $Q \subseteq N$ ask for the check and let us denote by x_Q the corresponding allocation of the cost C and eventually of the cost χ when the check is made. Consider the possible situations:*

a. *no agent asks for the check:* $x_\varnothing = (3, 3, 3, 3)$;
b. *agent 1 asks for the check and obtains it:* $x_{\{1\}} = (2, 0, 6, 6)$;
c. *agent 2 asks for the check and obtains it:* $x_{\{2\}} = (0, 2, 6, 6)$;
d. *agents 1 and 2 ask for the check and obtain it:* $x_{\{1,2\}} = (1, 1, 6, 6)$.

Comparing cases b., c. and d., the non-users may have an advantage from not asking for the check, with the risk that the exit is a., with an unfair solution.

A simple way to avoid the free-riding behavior is the following.

Mechanism 2. *Only who asks for the check pays the equal share of the check cost.*

Remark 1. *Applying Mechanism 2, the non-users that do not ask for the check pay a quota of the maintenance cost, whose amount is $\frac{C}{n}$ if the check is not obtained and $\frac{C}{|N \setminus Q|}$ if it is obtained; on the other hand, Mechanism 1 guarantees that non-users that ask for the check do not pay more than $\frac{C}{n}$. Under Mechanisms 1 and 2, strategy $\mathcal{A}$ is weakly dominant for non-users, even if they ignore the number m of non-users and the cost χ.*

Applying Mechanism 2 to the previous example, we obtain the following situations.

Example 3 (Example 2, with Mechanism 2). *The new allocations in cases b and c are* $x_{\{1\}} = (2, 4, 4, 4)$ *and* $x_{\{2\}} = (4, 2, 4, 4)$*, respectively, i.e. the non-user that does not ask for the check pays more than without the check.*

Example 4 (Example 2 revised). *Suppose that* $\chi = 7$*; in this case* $q = 3$*, i.e. the non-users cannot obtain the check and the solution is anyhow* $x = (3, 3, 3, 3)$*.*

Nevertheless, it is possible to extend the situations in which the check is made. The idea is to take into account the level of usage of the facility, supposing that it is available after the check.

4.2 Intensity of Usage

By Mechanisms 1 and 2 we have solved free-riding problems, but there is no guarantee that the quorum is reached. Obviously there are unfair situations that we cannot remove, such as the game with few non-users versus an expensive check cost and a cheap maintenance cost (see Example 4). In order to further reduce free-riding behaviors and increase fairness (here we think of check as fairness), we assume that each agent knows how much s/he used the facility and may calculate how much s/he has to pay at all, i.e. considering both maintenance and check costs.

More precisely, we suppose that when the check is made a division rule γ : $\mathbb{R} \longrightarrow \mathbb{R}^n_{\geq}$, $\gamma(C) = (\gamma_1(C), \gamma_2(C), \ldots, \gamma_n(C))$, provides the non-negative amount $\gamma_i(C)$ assigned to agent $i \in N$[1]. Then, the agents who asked for the check pay an equal share of the check cost χ plus the quota assigned by the division rule γ, while the other agents equally share the remaining part of the cost C, according to Mechanism 2.

We require that a fair division rule γ satisfies the following properties:

- *efficiency,* $\sum_{i \in N} \gamma_i = C$, i.e. the whole cost C is assigned;
- *weak monotonicity w.r.t. the level of usage,* i.e. whenever the level of usage of the facility is strictly higher for agent i than for agent j then $\gamma_i \geq \gamma_j$ for each C;
- *equal treatment of equal users,* i.e. whenever the level of usage of the facility is the same for agents i and j then $\gamma_i = \gamma_j$ for each C;
- *weak monotonicity w.r.t. the cost* C, i.e. given two different situations (e.g. two different periods) whose costs are C and C', respectively, with $C > C'$, then $\gamma_i(C) \geq \gamma_i(C')$ for every $i \in N$.

Remark 2

- *The equal division of* C *satisfies the properties above, i.e. the following approach generalizes the previous models.*

---

[1] When no confusion arises, we write γ instead of $\gamma(C)$.

- *No agent receives money for using the facility, whatever the level of usage.*
- *We do not require that $\gamma_i = 0$ if agent i does not use the facility; this may represent a fee for having the possibility of using the facility, even if the agent presently did not used it.*

Of course, there exist other fair division rules; on the other hand, the function γ should be not too much complex, in such a way that the agents may calculate the amount that they have to pay if the check is made. The underlying idea is that we enlarge the information available to the agents, so that also the users with a low level of usage that will be charged with a low share of the maintenance costs may become interested in asking for the check. For instance, we may think of assigning different slots of intensity of usage, e.g. low usage (up to a first threshold), medium usage (up to a larger threshold), high usage (over the larger threshold) and sharing the costs according to the slots. Another intuitive division rule assigns to the agents an amount proportional to the actual usage, e.g. if the facility is a printer each agent pays an amount proportional to the number of pages s/he printed. More sophisticated rules can be used, possibly accounting also the right of usage.

We stress that, according to Mechanism 2, only the agents that asked for the check may profit of it, otherwise the free-riding behavior of a user at low level may show up again; it will be made clearer in Example 5.

The division rule γ makes the request for the check non-disadvantageous for agent $i \in N \setminus M$ if

$$\frac{\chi}{|M \cup \{i\}|} + \gamma_i \le \frac{C}{n}. \tag{1}$$

Condition (1) descends from the following reasoning: after the introduction of Mechanisms 1 and 2 all the agents in the set M ask for the check, so if agent $i \in N \setminus M$ asks for the check then the cost χ is divided among at least $|M| + 1$ agents (other users may have the same interest in asking for the check but the information on the level of usage could not be common knowledge, differently from the information on who the users are), so if adding the individual quota of the maintenance cost γ_i the total is not larger than the equal sharing of the maintenance cost, it is convenient to ask for the check. In other words, the intensity of usage makes strategy $\mathcal{N}$ no longer dominant for the users.

Example 5. *Let $N = \{1, 2, 3, 4, 5\}, M = \{1, 2\}, C = 30, \chi = 11$; in this case $q = 2$, so the two agents in M obtain the check, and the final allocation is $x_{\{1,2\}} = (0, 0, 10, 10, 10)$. Now, suppose that agent 3 uses the facility at a low level, agent 4 uses the facility at a medium level, and agent 5 uses the facility at a high level; let the division rule of the cost C be such that $\gamma(C) = (0, 0, 1, 6, 23)$. In this case agent 3 satisfies Condition (1), so s/he may have an advantage from the check even if s/he is a user; in fact, asking for the check s/he pays at most $4.666 = \frac{11}{3} + 1$; according to Mechanism 2 the remaining part of the cost C is equally shared among agents 4 and 5, so the final allocation is $x_{\{1,2,3\}} = (3.666, 3.666, 4.666, 14.5, 14.5)$, i.e. all the agents that ask for the check profit from it.*

Remark 3. *Note that in Example 5 agent 4 does not satisfies Condition (1) as $\frac{\chi}{3} + \gamma_4 > \frac{C}{5}$; on the other hand, when $Q = \{1, 2\}$ agent 4 pays 10, that is more than what s/he would have paid asking for the check jointly with agents 1 and 2, i.e. $9.666 = \frac{11}{3} + 6$.*

In view of the previous remark, we can revise Condition (1) as follows.

The division rule γ makes the request for the check non-disadvantageous for agent $i \in N \setminus M$ if

$$\frac{\chi}{|M \cup \{i\}|} + \gamma_i \leq \frac{C}{|N \setminus M|} \text{ and } q \leq |M|. \tag{2}$$

Differently from Condition (1), in the case of Condition (2) we compare two more homogeneous situations; in fact, suppose that all the agents in M and another agent $i \in N \setminus M$ ask for the check obtaining it, then the left hand side corresponds to the cost for agent $i \in N \setminus M$ when all the agents in M and her/himself ask for the check obtaining it (if other users ask for the check, the cost for agent i may be lower), while the right hand side corresponds to the cost for agent $i \in N \setminus M$ when all the agents in M ask for the check obtaining it, so that the cost C is equally shared only among the users (if other users ask for the check, the cost for agent i may be higher).

Example 6 (Example 5, with Condition (2)). *Agents 1 and 2 are non-users, and agents 3 and 4 satisfy Condition (2), so all of them ask for the check and the allocation is $x_{\{1,2,3,4\}} = (2.75, 2.75, 3.75, 8.75, 23)$. Note that, according to Condition (2), agents 3 and 4 expected to pay 4.666 and 9.666, respectively.*

We remark that the agents have the information for computing the value of q and compare it with the cardinality of the set of non-users M, but they may be not aware if the quorum is reached when it is larger than $|M|$ and/or other agents in $N \setminus M$ asked for the check.

5   Concluding Remarks

In this paper, we reconsidered the problem in [1] focusing our attention on the free-riding behaviors that users and non-users may have. The free-riding of the users is faced in [1] via the naming procedure, offering the non-users the possibility of a check that certify who the users are. This procedure does not completely eliminates the free-riding of the users because the non-users may be afraid that they may pay more if the cost of the check is high and there are not enough requests so that the cost is shared among few agents; moreover, it is possible that non-users may behave as free-riders as they may not ask for the check with the hope that others non-users ask and pay for the check. Two suitable mechanisms were introduced in order to make the strategy "ask for the check" weakly dominant for the non-users. Then, we considered the possibility that the check provides also information on the level at which the facility is used by each agent. In this way, the strategy "not to ask for the check" is no longer dominant for those agents whose level of usage is sufficiently low; consequently, there is a larger number of situations in which the check is made, increasing the fairness of the cost sharing.

We want to remark that the enlargement of the possibilities of check may induce a new behavior for the users; in fact, since the total payment is $C + \chi$ with the check and C otherwise, the users at high level may propose to the other agents a kind of manipulation of the division of the maintenance cost, in such a way that it results advantageous for all the agents, especially when the cost of the check is high.

Of course, our proposal may be improved because the situation could be more complex, as in the following example.

Example 7 (Example 5 revised). *Suppose that* $\chi = 16$, *so the solution is* $x_{\{1,2,3\}} = (5.333, 5.333, 6.333, 14.5, 14.5)$, *i.e. agent 3 pays more asking for the check than otherwise. On the other hand, it is easy to notice that the first three agents pay in total* $17 = \chi + \gamma_3 = 16 + 1$ *if the check is made and* $18 = \frac{C}{n} \cdot 3 = 6 \cdot 3$ *otherwise. So, it is advantageous to ask for the check but a division of the cost* χ *different from the simple equal share among the agents that ask for the check is necessary.*

This is a classical cost allocation problem that can be solved via a cooperative game with transferable utility; we will consider this new setting in a forthcoming research.

Acknowledgment. The authors gratefully acknowledge the participants to the second workshop "Quantitative methods of group decision making" held at the Wroclaw School of Banking in November 2016 for useful discussions.

References

1. Briata, F.: Noncooperative games from TU games with information cost. Int. Game Theor. Rev. **13**, 301–323 (2011)
2. Hurwicz, L.: The design of mechanisms for resource allocation. Am. Econ. Rev. **63**, 1–30 (1973)
3. Maskin, E.: Mechanism design: how to implement social goals. Les Prix Nobel **2007**, 296–307 (2008)
4. Moretti, S., Patrone, F.: Cost allocation games with information costs. Math. Methods Oper. Res. **59**, 419–434 (2004)
5. Myerson, R.B., Satterthwaite, M.A.: Efficient mechanisms for bilateral trading. J. Econ. Theor. **29**, 265–281 (1983)
6. Nash, J.F.: Equilibrium points in n-person games. Proc. Natl. Acad. Sci. U.S.A. **36**, 48–49 (1950)
7. Roth, A.E.: The economist as engineer: game theory, experimentation, and computation as tools for design economics. Econometrica **70**, 1341–1378 (2002)
8. Vickrey, W.: Counterspeculation, auctions, and competitive sealed tenders. J. Finance **16**, 8–37 (1961)

Simulating Crowd Evacuation with Socio-Cultural, Cognitive, and Emotional Elements

C. Natalie van der Wal[1]([⊠]), Daniel Formolo[1], Mark A. Robinson[2], Michael Minkov[3], and Tibor Bosse[1]

[1] Department of Computer Science, Vrije Universiteit Amsterdam,
Amsterdam, Netherlands
c.n.vander.wal@vu.nl
[2] Socio-Technical Centre, Leeds University Business School, Leeds, UK
[3] Varna University of Management, Sofia, Bulgaria

Abstract. In this research, the effects of culture, cognitions, and emotions on crisis management and prevention are analysed. An agent-based crowd evacuation simulation model was created, named IMPACT, to study the evacuation process from a transport hub. To extend previous research, various socio-cultural, cognitive, and emotional factors were modelled, including: language, gender, familiarity with the environment, emotional contagion, prosocial behaviour, falls, group decision making, and compliance. The IMPACT model was validated against data from an evacuation drill using the existing EXODUS evacuation model. Results show that on all measures, the IMPACT model is within or close to the prescribed boundaries, thereby establishing its validity. Structured simulations with the validated model revealed important findings, including: the effect of doors as bottlenecks, social contagion speeding up evacuation time, falling behaviour not affecting evacuation time significantly, and travelling in groups being more beneficial for evacuation time than travelling alone. This research has important practical applications for crowd management professionals, including transport hub operators, first responders, and risk assessors.

Keywords: Crowd behaviour · Crowd management · Crowd simulation · Evacuation · Emotional contagion · Social dynamics · Culture · Cognition · Group-decision making

1 Introduction

Crisis management and prevention involves preparing for many different emergency situations. This research focuses on studying the socio-cultural, cognitive, and emotional factors influencing an evacuation from a building, such as a transport hub. This is important, because few crisis managers and risk assessment professionals currently deal with these factors and their resulting behaviours. Accordingly, this research developed and validated a crowd evacuation simulation model that includes socio-cultural, cognitive, and emotional factors in order to simulate what-if scenarios. Consequently, it

© Springer International Publishing AG 2017
N.T. Nguyen et al. (Eds.): TCCI XXVII, LNCS 10480, pp. 139–177, 2017.
https://doi.org/10.1007/978-3-319-70647-4_11

will help transport hub operators, crisis managers, risk assessment professionals, and policy makers understand human behaviour, deal with socio-cultural crowd diversity, and ultimately save lives.

Faster evacuation from public buildings during emergencies saves more lives. Observations of actual emergencies show that people tend to be slow to respond to evacuation alarms (taking up to 10 min) and take the familiar route out instead of the nearest exit [4, 7, 14, 21, 23, 30]. These risky behaviours stem from being unfamiliar with the environment, not seeing immediate signs of danger, and following others' (unsafe) behaviour, leading to preventable deaths in many disasters. For instance, in the Station Nightclub fire, in Rhode Island in 2003, the majority of people tried to escape back through the familiar main entrance, leading to falls, crushing, and 100 deaths. Many of the 56 deaths in the Bradford City Stadium fire in 1985 could have been prevented if response time to the fire had been faster [3], and similarly slow responses were found among occupants of the World Trade Center towers during the 9/11 terror attacks in New York City [23]. In recent emergencies, some people have even remained in dangerous areas to film events with their smartphones instead of escaping (Nice Boulevard, 14/07/2016; Westgate Shopping Centre, Nairobi, 21/9/2013).

Current crowd evacuation models simulate how crowds move through built environments [9], enabling ethical tests of how to improve crowd movements in emergency evacuations. In addition to informing how to build safer buildings, computer models can identify safer behaviours in existing buildings. For example, it is well-documented that not running leads to faster evacuations due to fewer falls and less congestion at the exit [17, 36]. However, traditional computer models of evacuations have been criticized for being unrealistic, because they treat people as 'moving particles' with identical characteristics [9, 36]. Such models wrongly assume that all people will respond to alarms without delay, know their way, and take the nearest exit. As noted above, however, each of these assumptions has been proven wrong [4, 7, 14, 21, 23, 30].

The aim of this research, therefore, is to develop and validate an evacuation simulation model that includes socio-cultural, cognitive, and emotional factors, to address the need for crowd models to incorporate more realistic human behaviours. To do so, the model developed here draws on insights from social and cross-cultural psychology, interviews with crisis management experts, and is based on scientific findings and literature. Furthermore, the model is validated against data from an evacuation drill related to the existing EXODUS evacuation model [13, 26]. It is intended that this model will help transport hub operators, crisis managers, risk assessment professionals, and policy makers understand human behaviour, deal with socio-cultural crowd diversity, and ultimately save lives.

The paper is organised as follows. First, the background literature on crowd evacuation models is reviewed and the current approach is introduced in Sect. 1.1. In Sect. 2, the formal model is presented, followed by the validation and simulation results in Sect. 3. The work is then summarised and discussed in Sect. 4.

1.1 Background Evacuation Models

There are many different approaches for crowd evacuation simulations, of which Zheng et al. [48] describe seven: (1) cellular automata, (2) lattice gas, (3) social force, (4) fluid dynamics, (5) agent-based, (6) game theory, and (7) animal experiments. In microscopic models (e.g. cellular automata, lattice gas, social force, agent-based models), the pedestrian is modelled as a particle. However, in macroscopic models (e.g. fluid dynamic models), a crowd of pedestrians is modelled as a fluid. In conclusion, Zheng et al. [48] concluded that in further research, evacuation models should: (1) combine different approaches, and (2) incorporate psychological and physiological elements. Our IMPACT model addresses both of these recommendations.

Moreover, Templeton et al. [39] conclude that current crowd simulations do not include psychological factors and therefore cannot accurately simulate the collective behaviour that has been found in extensive empirical research on crowd events. Specifically, they argue that crowd members should be able to identify with other people in crowd simulations to form psychological sub-groups known as in-groups. This is critical for evacuation models, as research indicates that people are more likely to help fellow in-group members during emergencies [8]. Accordingly, our IMPACT model also incorporates social identity.

Most of the evacuation models that Santos and Aguirre [36] reviewed do not model social dimensions, such as group decision making, but focus more on physical constraints and factors such as walking speed, walkways, and stairways, to find the optimal crowd flow for the evacuation process. Agents are rational in these simulations: they can find the optimal escape route, avoid physical obstructions and, in some models, even overtake another person obstructing them. However, even though these models do include parameters like gender, age, individual walking speeds, and different body dimensions, they still lack socially interactive characteristics such as the monitoring of others. Again, to address this, our IMPACT model incorporates such social processes.

Santos and Aguirre [36] also reviewed the incorporation of social and psychological factors into evacuation simulation models, noting their inclusion in three models: (1) FIRESCAP, (2) EXODUS, and (3) Multi-Agent Simulation for Crisis Management (MASCM). EXODUS includes 22 social psychological attributes and characteristics for each agent, including age, sex, running speed, dead/alive, and familiarity with the building. Agents can also perform tasks before evacuating the building, such as picking up a purse or searching for a lost child. Still, the agents in EXODUS cannot have micro-level social interactions that would create a collective understanding of the situation for the group. However, MASCM does include social interaction with so-called 'evacuation leaders' who can communicate ('please follow me') and start to walk along the evacuation route, or find an evacuee, or wait for an evacuee to approach them. Finally, FIRESCAP implements the social theory of 'collective flight from a perceived threat'. The egress is a result of a socially-structured decision making process guided by norms, roles, and role relations.

From this literature review, it can be concluded that the ideal simulation approach for realistic crowd evacuation models should seek to develop sub-models that include an active, 'investigative', socially-embedded agent that assesses the state of other people and defines the situation collaboratively. Essentially, then, group dynamics must be considered, and our IMPACT model aims to address this.

1.2   Current Approach

Based on the lack of psychological and socio-cultural factors in existing evacuation models, we created our IMPACT evacuation model based on an earlier model called ASCRIBE [2]. This allows for the social contagion of emotional and mental states, and enables group decision making and other social dynamics [1, 2]. The ASCRIBE model has outperformed other models in reproducing real crowd panic scenes and was extended here with many psychological and socio-cultural factors – such as familiarity, falls, and prosocial behaviour – and applied to a specific evacuation scenario [41]. The evacuation dynamics were modelled using agent-based belief-desire-intention (BDI) and network-oriented modelling approaches [32, 40]. A first version of the IMPACT model was introduced in [43] and the further-developed and validated model was introduced in [12]. The final version of the IMPACT model presented here has now been fully refined and certain characteristics have been updated. We introduce it here with its most important findings. The updates concern speed, falls, compliance levels, egress flowrate, observation distance, helping behaviour, and cultural divisions, and these are based on psychological and socio-cultural research as described below.

1.3   Background Psychological and Socio-Cultural Factors
in the IMPACT Model

Overview. Although the computer simulation of crowd behaviour has been ongoing for several decades, most existing models are still founded on erroneous assumptions of human behaviour and movement as linear, logical, and driven primarily by the laws of physics [4]. A key reason for this has been the disciplinary division in crowd behaviour research. Modellers engaged in crowd simulation are typically drawn from technical fields, such as computer science and engineering, while psychologists and other social scientists who study crowd behaviour do not generally use computer simulation methods [18]. Consequently, only truly interdisciplinary research can effectively simulate crowd behaviour, particularly in emergencies, in complex systems comprising both social and technical elements [5]. To address these issues in our IMPACT model, alongside the conventional features of traditional crowd simulation models we have included additional psychological and socio-cultural elements. For instance, at an individual level, we have simulated the effect of people's socio-cultural characteristics such as age, gender, and nationality on their behaviour (e.g. based on the national cultural clusters in [35]) in emergencies; while, at a group level, we have simulated social processes such as social identity [8] and emotional contagion [1, 2].

Speed. The walking speeds varied for each demographic group (children, adult males, adult females, elderly males, elderly females) and were based on the observational work of Willis et al. [46], ranging from 1.12 m/s to 1.58 m/s. We calculated running speeds by multiplying the walking speed for each demographic group by three – to account for the luggage, belongings, and clothes that people wear while travelling – to yield speeds between 3.36 m/s and 4.75 m/s. Moreover, a crowd congestion factor was added that reduces the speed according to the number of agents within the same square metre: $\leq$ 4 people (no speed reduction), 5 people (62.5% reduction), 6 people (75%),

7 people (82.5%), 8 people (95%). These speed adjustments were based on research by Still [38], where 8 is the maximum number of people per square metre and 4 the number of people at which speed reduces.

Falls. The number of falls in the initial model seemed unrealistically high during structured simulations. So, we manually tuned the value to a more realistic level by visually inspecting the movement patterns during many different settings. This resulted in a new rule: if there are more than 4 people in the same square metre as the agent and if he is running faster than 3 m/s, then there is a 5% chance of a fall for each new movement.

Compliance. In the current version, the probability of compliance is based on data from Reininger et al.'s [33] study of gender differences in hurricane evacuation, modified for different age groups using data from Soto et al.'s [37] personality study. The model has 6 compliance values according to the category of the agent: male or female, and child, adult, or elderly. The precise levels can be found in Sect. 2.

Egress flowrate at each exit. The maximum flowrate is 6 people per exit per second (p/m/s), based on guidelines from Still [38] indicating an egress flowrate of 82 people/metre/minute (p/m/m), equivalent to 1.37 p/m/s, then multiplied by 4 (as doors are 4 m wide) to indicate 5.47 people per exit door per second.

Observation distance. Public distance (space in which social interactions are still possible, extending the personal and formal social interaction space) is 12–25 feet (3.7–7.6 m), in relation to public speaking to large groups, while no social interaction is possible over 25 feet [15], though this might not take shouting into account. Considering the size of the environment that was implemented in the model (e.g. a square room of 20 × 20 m), it was decided to keep the observation distance (i.e. the maximum distance at which staff instructions could be understood) at 5 m rather than 10. Otherwise, at 10 m, the passengers could observe everything in the building from the centre and the important effects of social contagion would be downplayed in the simulations.

Helping. The probabilities of helping others during the emergency evacuation were modelled as a function of the characteristics of helpers and fallers. This was based on research indicating that, in emergencies: (a) men are most likely to help others, (b) women, children, and older adults are most likely to receive help [10], and (c) people are more likely to help members with a shared identity [8]. The precise probabilities can be found in Sect. 2.1.

Culture. In the model, the passengers are divided into different clusters of culturally similar nationalities based on previous research [35]. Data concerning the percentage of English speakers for each country in each cluster were then obtained, where available, from multiple verified and official sources compiled by Wikipedia [45]. We then calculated a weighted average percentage of English speakers in each cluster – using the population sizes of each cluster's constituent countries – and these were the values used in the simulation model to determine the percentage of passengers from each cluster who could understand an English instruction by a staff member or public announcement. The precise probabilities can be found in Sect. 2.1.

Group decision making. Like in previous work [1, 2], group decision making is based on findings from social neuroscience to make a biologically plausible human-like model. Decision making is modelled as both an individual process called somatic marking and a social group process based on mirroring of cognitive and emotional states [6, 34]. Damasio's somatic marking hypothesis is a theory of decision making which provides a central role to emotions felt [6]. Each decision option induces a feeling to mark that option. In social neuroscience, neural mechanisms have been discovered that account for mutual mirroring effects between mental states of different people. For example, when one expresses an emotion in a smile, another person can observe this smile which automatically triggers preparation neurons (called mirror neurons) for smiling within this other person and consequently generates the same emotion. Similarly, mirroring of intentions and beliefs can be considered. This is called emotional contagion (for emotions alone) or social contagion (for emotional and mental states) in this work.

2   Model

2.1   Formal Model

Figure 1 gives an overview of the formal model, showing the four modules of each passenger and how they interact. The passenger has individual characteristics – such as age, gender, familiarity, and group membership – which influence their interactions. For example, familiarity influences the choice of exit (people-environment interaction), while age, gender, and group membership influence the pro-social behaviour (people-people interactions). The full details of these four modules, their constituent

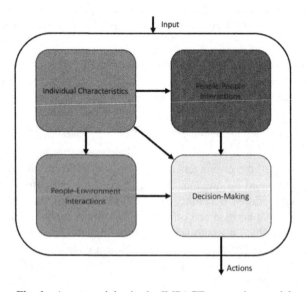

Fig. 1. Agent modules in the IMPACT evacuation model

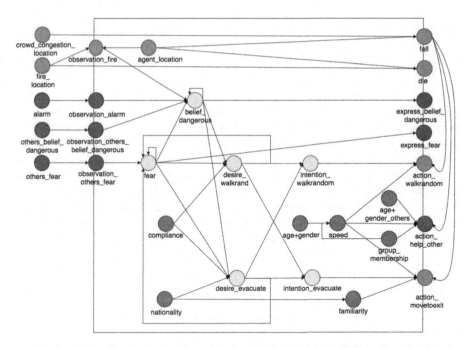

Fig. 2. Dynamic relationships between concepts in the IMPACT evacuation model

concepts, and their dynamic relationships are shown in Fig. 2, using the same coloured key as Fig. 1 for the modules.

Below, all the formal rules of the proposed model are presented in the form of mathematical formulas representing all dynamic relationships between all concepts from Fig. 2. Creating the formal model in this way, using mainly difference equations, is based on the network oriented modelling approach [40].

Firstly, the following environmental states have the value 0 ('off') or 1 ('on'). These are 'inputs' of the model and vary over time. For example, the fire_alarm is 'on' after three minutes of the simulation and the public_announcement is 'on' one minute after the fire_alarm is 'on'.

$$\text{crowd\_congestion\_location}(t); \ \text{fire\_location}(t); \ \text{alarm}(t); \ \text{staff\_instructions}(t);$$
$$\text{public\_announcement}(t) \tag{1}$$

The aggregated impacts of others on agent x, for the levels of the belief that the situation is dangerous and the levels of fear, are calculated as a weighted sum at every time step, based on previous work [1, 2]:

$$\text{others\_belief\_dangerous}_x(t) = ssum_\lambda(\omega_{y1x} \cdot belief\_dangerous_{y1}, \ldots, \omega_{kx} \cdot$$
$$belief\_dangerous_k) = ssum_\lambda(\omega_{y1x} \cdot belief\_dangerous_{y1} + \ldots + \omega_{kx} \cdot$$
$$belief\_dangerous_k) = \frac{\sum_k^{y1} \omega_{y1x} \cdot belief\_dangerous_{y1}(t)}{\sum_k^{y1} \omega_{y1x}}. \tag{2}$$

$$
\text{others\_fear}_x(t) = ssum_\lambda\left(\omega_{y1x}\cdot fear_{y1},\ldots,\omega_{kx}\cdot fear_k\right) = ssum_\lambda\left(\omega_{y1x}\cdot fear_{y1}+\ldots+\right.
$$
$$
\left.\omega_{kx}\cdot fear_k\right) = \frac{\sum_k^{y1}\omega_{y1x}\cdot fear_{y1}(t)}{\sum_k^{y1}\omega_{y1x}}. \tag{3}
$$

whereby $\lambda = \sum_k^{y1}\omega_{y1x}$

All observations of events or other passengers are calculated as stated below. The observation_fire becomes 1 if the passenger is within a distance of 5 m, representing the observation distance which is adjustable by the modeller, based on [15], see Sect. 1.3. When the fire alarm sounds, then 50% of the time the passenger will observe this alarm and this, in turn, will change the passenger's belief_dangerous to 1. This represents the risk-taking passengers have, as not all passengers react quickly to a fire alarm [21, 23, 30]. Note that, for example, for observation_others_fear(t) = others_fear(t) a simplification of the real world has been made to model the values to match each other instantaneously instead of with a delay, as further detail was not necessary in the model.

$$
\text{observation\_fire}(t) = 1 \text{ if } \left(\sqrt{(x_1-x_2)^2+(y_1-y_2)^2}\le 5\right) \text{ else } 0; \text{ where by agent\_location}(t) = \tag{4}
$$
$(x_1\,y_2)$ and fire_location$(t) = (x_2\,y_2)$.

$$
P(\text{observation\_alarm}(t) = 1 \mid \text{alarm}(t) = 1) = 0.5. \tag{5}
$$

$$
\begin{aligned}
\text{observation\_others\_belief\_dangerous}(t) &= \text{others\_belief\_dangerous}(t);\\
\text{observation\_others\_fear}(t) &= \text{others\_fear}(t);\\
\text{observation\_staff\_instr}(t) &= \text{staff\_instructions}(t);\\
\text{observation\_pa}(t) &= \text{public\_announcement}(t)
\end{aligned} \tag{6}
$$

If there is a fire at the same location as the passenger, then the passenger dies. Die(t) has a binary value of 0 ('not dead') or 1 ('dead'). This strict rule was chosen as more detail was not necessary for the goal of this model. We chose not to model the effect of the fire and smoke, like the heat and toxicity in the room, so we could purely focus on the human behavioural effects in the simulations not combined with the effects of the fire.

$$
\text{die}(t) = 1(\text{if fire\_location} == \text{agent\_location}) \text{ else } 0. \tag{7}
$$

Each passenger has an initial speed based on his/her age and gender, based on [38, 46], see Sect. 1.3.

At $t = 0$:

- If age + gender = female adult then basic speed = $0.9 +$ rand $(0, 0.5)$.
- If age + gender = male adult then basic speed $= 1 +$ rand $(0, 0.5)$.
- If age + gender = child then basic speed = $0.5 +$ rand $(0, 0.5)$.
- If age + gender = female elderly then basic speed = $0.9 +$ rand $(0, 0.5)$.
- If age + gender = male elderly then basic speed = $0.9 +$ rand $(0, 0.5)$.
- If group_membership = 1, then speed = min(basic speeds of other members)

$+ 0.4 \cdot (\max(\text{basic speeds of other members}) - \min(\text{basic speeds of other members}).$
- If group membership $= 0$, then speed $=$ basic speed.   (8)

Whereby: rand is a random number, min = minimum, and max = maximum.

Each passenger has an initial compliance level based on his/her age and gender, based on [33, 37], see Sect. 1.3.

At t = 0:

- If age + gender $=$ male child then compliance $= 0.89$.
- If age + gender $=$ female child then compliance $= 0.89$.
- If age + gender $=$ male adult then compliance $= 0.89$.
- If age + gender $=$ female adult then compliance $= 0.94$.   (9)
- If age + gender $=$ male elderly then compliance $= 0.92$.
- If age + gender $=$ female elderly then compliance $= 0.97$.

Each passenger has a 5% chance (i.e., a 0.05 probability) of falling when there is crowd congestion at their location, as explained in Sect. 1.3. Fall(t) has a binary value of 0 ('not fallen') or 1 ('fallen').

$$P(\text{fall}(t) = 1 | \text{crowd\_congestion\_location} == \text{agent\_location}) = 0.05. \qquad (10)$$

Each passenger has a belief about how dangerous the situation is. This belief has a value between 0 ('minimum danger') and 1 ('maximum danger'). The belief will increase to 1 when a fire or alarm is sensed. The beliefs of other passengers can decrease or increase the passenger's own belief, based on mirroring/contagion mechanisms as described in Sect. 1.3, based on previous research [1, 2]. The passenger's fear level influences his belief (somatic marking): if the amount of fear is higher than the belief, it will increase the belief, and if the amount of fear is lower than the belief, it will decrease the belief. The belief is also based on the passenger's belief from the previous time-step (persistence). The equations are presented in both difference and differential equation format to show how, hereafter, every difference equation can be translated into a differential equation.

$belief\_dangerous(t + \Delta t) = belief\_dangerous(t) + \eta \cdot (\max(\omega_{sensing} \cdot fire(t), \omega_{sensing} \cdot$
$alarm(t), \omega_{persisting} \cdot belief\_dangerous(t), sum\left(\frac{\omega_{affectivebiasing} \cdot fear(t) + aggbeliefs_x(t)}{\omega_{affectivebiasing} + 1}\right)) -$
$belief\_dangerous(t)) \cdot \Delta t.$

$$(11)$$

$\frac{dbelief\_dangerous}{dt} = \eta \cdot (\max(\omega_{sensing} \cdot fire(t), \omega_{sensing} \cdot alarm(t), \omega_{persisting} \cdot$
$belief\_dangerous(t), sum\left(\frac{\omega_{affectivebiasing} \cdot fear(t) + aggbeliefs_x(t)}{\omega_{affectivebiasing} + 1}\right) - belief\_dangerous(t))$

$$(12)$$

whereby, $aggbeliefs_x(t) = ssum_\lambda(\omega_{y1x} \cdot belief\_dangerous_{y1}(t), \ldots, \omega_{kx} \cdot$
$belief\_dangerous_k(t)) = ssum_\lambda(\omega_{y1x} \cdot belief\_dangerous_{y1}(t) + \ldots + \omega_{kx} \cdot$
$belief\_dangerous_k(t)) = \frac{\sum_k^{y1} \omega_{y1x} \cdot belief\_dangerous_{y1}(t)}{\sum_k^{y1} \omega_{y1x}}.$
$\lambda = \sum_k^{y1} \omega_{y1x}$

The amount of fear a passenger feels is based on the fear level of the previous time-step (persistence), the levels of intentions to evacuate (amplifying fear) or walk randomly (decreasing fear), the other passengers' levels of fear (emotional contagion), and the staff instructions or public announcements they observe (decreasing fear). These processes are based on mirroring/contagion mechanisms as described in Sect. 1.3, based on previous research [1, 2]. The fear value ranges from a minimum of 0 ('no fear') to a maximum of 1 ('maximum fear').

$$\begin{aligned} fear(t + \Delta t) = fear(t) + \eta \cdot (\max{} (\omega_{persisting} \cdot fear(t), alogistic(aggfears(t), \\ \omega_{amplifyingfeeling} \cdot desire_{evacuate(t)}, \omega_{inhibitingfeeling} \cdot desire_{walkrand(t)}, \omega_{decreasingfear} \\ \cdot observation_{staf_{instr(t)}}, \omega_{decreasingfear} \cdot observation_{pa(t)})) - fear(t)) \cdot \Delta t. \end{aligned} \tag{13}$$

whereby, aggfears(t) is calculated similarly as $aggbeliefs_x(t)$ (see Eq. 12) and $alogistic_{\sigma\tau}(V_1, \ldots, V_k) = (\frac{1}{1 + e^{-\sigma(V_1 + \ldots + V_k - \tau)}}) - \frac{1}{1 + e^{\sigma\tau}})(1 + e^{-\sigma\tau}).$

The desire to evacuate value ranges from 0 ('minimal desire') to 1 ('maximal desire'). It is amplified by the level of compliance, the passenger's belief of how dangerous the situation is (cognitive responding), the passenger's level of fear (somatic marking), and staff instructions or public announcements to evacuate. The somatic marking and cognitive responding are processes based on mirroring/contagion mechanisms as described in Sect. 1.3, based on previous research [1, 2].

$$\begin{aligned} desire\_evacuate(t + \Delta t) = desire\_evacuate(t) + \eta \cdot ((compliance \cdot \\ (\max(\omega_{amplifyingevacuation} \cdot belief\_dangerous(t), \omega_{amplifyingevacuation} \cdot \\ fear(t), \omega_{amplifyingevacuation} \cdot observation\_staff\_instr(t), \omega_{amplifyingevacuation} \cdot \\ observation\_pa(t)))) - desire\_evacuate(t)) \cdot \Delta t. \end{aligned} \tag{14}$$

Whereby,

$$ssum_\lambda(\omega_1 \cdot V_1(t), \ldots, \omega_k \cdot V_k,) = ssum_\lambda(\omega_1 \cdot V_1(t), \ldots, \omega_k \cdot V_k,) = \frac{\sum_k^1 \omega_1 \cdot V_1(t)}{\sum_k^1 \omega_1}, \lambda = \sum_k^1 \omega_1.$$

The value of the desire to walk randomly ranges from 0 ('minimal desire') to 1 ('maximal desire'). It is inhibited by the level of compliance, the passenger's belief of how dangerous the situation is (cognitive responding), the passenger's level of fear (somatic marking), and staff instructions or public announcements to evacuate. The somatic marking and cognitive responding are processes based on mirroring/contagion mechanisms as described in Sect. 1.3, based on previous research [1, 2].

$$desire\_walkrand(t + \Delta t) = desire\_walkrand(t) + \eta \cdot (compliance \cdot (1-$$
$$\max(\omega_{inhibitingwalkrand} \cdot belief\_dangerous(t), \omega_{inhibitingwalkrand} \cdot$$
$$fear(t), \omega_{inhibitingwalkrand} \cdot observation\_staff\_instr(t), \omega_{inhibitingwalkrand} \cdot \qquad (15)$$
$$observation\_pa(t)) - desire\_walkrand(t)) \cdot \Delta t.$$

The intention to evacuate value ranges from 0 ('minimal intention') to 1 ('maximal intention'), and so too does the intention to walk randomly value. To decide whether the desire to evacuate or walk randomly is larger, a logistic function is used, and this outcome is then multiplied by the desire to walk randomly. This, in turn, is multiplied by (1-fall(t)) to make sure it is only a value larger than 0 when the passenger has not fallen. When the passenger has fallen, the value will become 0, then the passenger cannot actually walk randomly or evacuate.

$$intention\_evacuate(t + \Delta t) = intention\_evacuate(t) + \eta \cdot ((1 - fall(t)) \cdot$$
$$desire\_evacuate(t) \cdot logistic((\omega_{amplifyingintention} \cdot desire\_evacuate(t), \qquad (16)$$
$$\omega_{inhibitingintention} \cdot desire\_walkrand(t)) \cdot \Delta t.$$

$$intention\_walkrand(t + \Delta t) = intention\_walkrand(t) + \eta \cdot ((1 - fall(t)) \cdot$$
$$desire\_evacuate(t) \cdot logistic((\omega_{inhibitingintention} \cdot desire\_evacuate(t), \qquad (17)$$
$$\omega_{amplifyingintention} \cdot desire\_walkrand(t)) \cdot \Delta t.$$

whereby: $logistic_{\sigma,\tau}(V_1, \ldots, V_k) = \frac{1}{1+e^{-\sigma(V_1 + \ldots + V_k - \tau)}})$.

The action movetoexit is a combination of the speed of the passenger and his target (i.e. the location/exit he moves towards). The value of the intention to evacuate influences the speed of moving to the exit. The familiarity, observation of staff instructions, and the public announcement all influence the choice of exit [4, 14].

If (familiarity $=$ 1 OR observation_staffinstructions $=$ 1 OR observation_pa $=$ 1) then action_movetoexit(t) $=$ (target $=$ nearest exit) AND (speed $=$ intention_evacuate(t)· speed) else action_movetoexit(t) $=$ (target $=$ entrance) AND (speed $=$ intention_evacuate(t) · speed). $\qquad (18)$

The action walkrandom is a combination of the speed and heading of the agent in the environment. The value of intention_walkrand is multiplied by the maximum speed of the agent.

$$action\_walkrand(t) = (heading = random) \text{ AND } (intention\_walkrand \cdot speed). \quad (19)$$

The action help_other is calculated as stated below, based on previous research [8, 10], as described in Sect. 1.3.
When

$$\sqrt{(x_1 - x_2)^2 + (y_1 - y_2)^2} \leq 5. \qquad (20)$$

Table 1. Probabilities of helping a fallen passenger

Helper passenger	Social identity	Fallen passenger					
		Male child	Male adult	Male elderly	Female child	Female adult	Female elderly
Male adult	In-group	0.30	0.15	0.30	0.40	0.30	0.40
Male elderly	In-group	0.15	0.08	0.15	0.20	0.15	0.20
Male adult	Out-group	0.25	0.13	0.25	0.34	0.25	0.34
Male elderly	Out-group	0.13	0.06	0.13	0.17	0.13	0.17
Female adult	In-group	0.15	0.08	0.15	0.20	0.15	0.20
Female elderly	In-group	0.08	0.04	0.08	0.10	0.08	0.10
Female adult	Out-group	0.13	0.06	0.13	0.17	0.13	0.17
Female elderly	Out-group	0.06	0.03	0.06	0.08	0.06	0.08

whereby agent_location(t) = $(x_1\ y_2)$ and agent_location of other passenger(t) = $(x_2\ y_2)$ and other passenger fall(t) = 1, then the chance of helping depends on the age + gender of the helper and the fallen passenger and whether they share a social identity (in-group) or not (out-group). The overall probability of helping is shown in Table 1.

The expressions of fear and the passenger's belief of the situation are modelled in a simple way, where the values match each other instantaneously instead of with a delay, as further detail was not necessary in the model.

$$\text{express\_belief\_dangerous}(t) = \text{belief\_dangerous}(t); \text{express\_fear}(t) = \text{fear}(t) \quad (21)$$

2.2   Pseudo-code and Model Overview

The model was implemented in the NetLogo multi-agent language [25]. To do so, the formal model presented in the previous section was transformed into multiple IF THEN rules. An example of how these rules were translated into NetLogo code is shown below, taking Formula *18* (see previous section) as an example. It is shown that for each agent in the model the heading (direction) is set as a random number between 0 and 360 (degrees), and then based on the age and gender of the agent a speed is also set. Then, for the action to walk randomly, the level of the intention is multiplied by the speed.

```
;-- EXAMPLE RULE IN NETLOGO --
ask agents [
    set heading random 360
    if st_gender = 0 and st_age = 1 [set speed 0.9 +
random-float 0.52]    ;female adult
    set st_action_walkrandom st_intention_walkrand *
speed
]
```

Figure 3 shows the activity diagram of the created simulator focusing on the internal model. The system updates internal states and actions of each agent. After that, it updates the environment, considering the actions of the agents, and finalizes the cycle by updating the statistics. The simulation stops when all agents are either evacuated or dead. At any moment, the user can change the parameters available on the interface and influence the environment or agents.

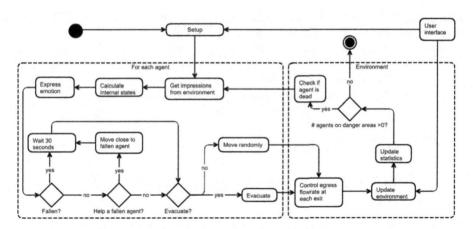

Fig. 3. Activity diagram overview of the IMPACT crowd evacuation model

3   Validation and Structured Simulation Results

3.1   Validation Results: IMPACT Model Versus EXODUS Benchmark

Our IMPACT model has been compared with a benchmark to establish its validity. In [12] the validation process and results have been explained and discussed already, and a summary is provided here. The EXODUS model [26] was selected as a benchmark

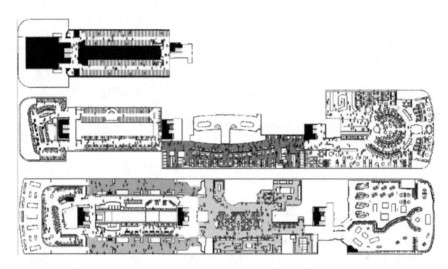

Fig. 4. Scenario of the software simulation.

for the IMPACT model, as it is accepted by specialists in this area as realistic [26]. The environment selected is called SGVDS1, a complex ship environment composed of three floors, with different escape routes to the four assembly areas [13] (Fig. 4).

A validation experiment was conducted comparing three versions of the IMPACT model with the benchmark of the EXODUS model (see Table 2 for the experimental design). The IMPACT model covers more aspects than the benchmark EXODUS model, however, so some of the IMPACT model's variables were fixed to enable a fair comparison:

Table 2. Results of the validation protocol for the overall arrival times.

Condition	Benchmark	Experimental condition 1	Experimental condition 2	Experimental condition 3
Explanation	Exodus SGVDS1 data	No Social Contagion. Response time and Speed taken from the benchmark	No Social Contagion. Response times and Speed calculated by the model itself	Social Contagion activated. Response times and Speed calculated by the model itself
FET	585 (s)	498.6 (s)	543.4 (s)	516.6 (s)
TAT	0	14.77	7.11	11.69
ERD	0	0.568171	0.575657	0.565754
EPC	0	0.724621	0.731295	0.731634
SC	0	0.522105	0.423135	0.451471

- Familiarity: it was assumed that everybody was not familiar with the environment.
- Relationship: it was assumed that all passengers were unrelated.
- Social contagion: this was 'on' or 'off', depending on the experimental condition (see Table 2).
- The passenger's speed: in experimental condition 1 the speeds indicated in [13] were used. In experimental conditions 2 and 3, the speed was calculated by the IMPACT model.
- Groups and Helping: these were not considered in any experimental condition.

The outcome measures of the validation experiment are: (1) Final Evacuation Time (FET); (2) the percentage difference between the predicted and Total Assembly Time (TAT); (3) the curve differences between the predicted and expected arrivals to the Assembly Areas (exits). This last measurement is calculated based on Euclidean Relative Difference (ERD), Euclidean Projection Coefficient (EPC), and Secant Cosine (SC). In [13] it is stated that a 'good' TAT should be below 40, which is true for all experimental conditions here. For ERD, all experimental conditions are over, but close to, the expected boundary that is ≤ 0.45, while for EPC, the results stay within the expected boundaries of $0.6 \leq EPC \leq 1.4$. For SC, the values are below the boundary 0.6, but close to the acceptance threshold. See Table 2 for all the results. In Fig. 5 below, the assembly curves of the benchmark and the three IMPACT versions (the three experimental conditions) are shown. These results show that on all measures, the IMPACT model is within or close to the prescribed boundaries, thereby establishing its validity.

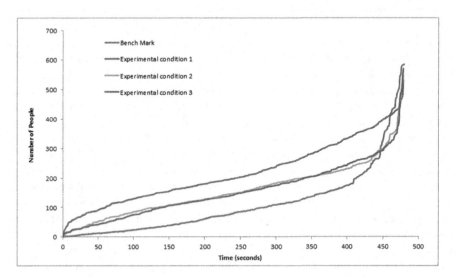

Fig. 5. Total arrival time pattern for one simulation run of EXODUS benchmark and IMPACT experimental conditions 1, 2 and 3.

3.2   Simulation Experiments Setup

Number of Repetitions. To determine the number of repetitions for each combination of factor and level, an evacuation scenario with the most variability was run 100 times. First, the cumulative averages and variances in evacuation time were inspected to detect the threshold number of repetitions at which evacuation time stabilised. Second, Eq. 22 below was used to find the minimum number of repetitions (56) to guarantee that the error in the outcome results is within 5% of the maximum error with a 95% confidence level. Then, 60 repetitions of each variation were run and the results presented in this Section represent the average of these 60 runs.

$$n \geq [100 \cdot Z \cdot s/r \cdot \bar{x}]^2 = 56.61599 \rightarrow 60 \, \text{samples} \qquad (22)$$

Whereby,

$Z = $ *confidence interval of* 95%; $s = $ *standard deviation,* 53.4287
$r = $ *maximum error of* 5% $\bar{x} = $ *evacuation time average of* 100 *samples*

Outcome measures and emergence. There are three outcome measures for each simulation experiment: (1) evacuation time, (2) total falls, and (3) response time. The evacuation time was measured as the number of seconds from the onset of the fire until all (living) passengers have evacuated. The number of falls was measured cumulatively (all falls in total in one simulation run). The individual response time was measured as the time between the onset of the fire until the passenger develops the intention to move to the exit. The reported response time is the average of all individual's response times.

Besides these outcome measures, emergence is of interest in the analyses. Emergence is the spontaneous establishment of a qualitatively new behaviour through non-linear interactions of many objects or subjects [17]. In other words, it is a behaviour observed at the group level, which cannot be directly explained from the individual behavioural rules. This could lead to unexpected findings in our simulation experiments, because the hypotheses are formulated based on individual behavioural rules, since a priori you do not always know what group level behaviour will occur. There are important crowd movement phenomena related to evacuation situations known from the literature, such as herding, the faster-is-slower-effect, and collective intelligence [16, 17]. Herding refers to a situation that is unclear and causes individuals to follow each other instead of taking the optimal route [16]. The faster-is-slower-effect refers to when, in evacuation situations, certain processes take longer at high speed; so, waiting can sometimes help competing people (competing for space) and speed up the average progress [17]. Collective intelligence, as Helbing and Johansson name it, is emergent functional behaviour of a large number of people resulting from interactions of individuals instead of individual reasoning [17].

We hope our model will create these emergent phenomena, as that would prove our model can create self-organisation [9]. Self-organisation can be defined as the spontaneous establishment of qualitatively new behaviour through non-linear interaction of many objects or subjects without the intervention of external influences. However, we do not expect our model to show emergent lane formation and the zipper effect [9]. Lane formation is a process where a number of lanes of varying width form dynamically at a corner; however, the passengers in our model do not have to go around a corner towards the exit.

Other evacuation modellers have studied behavioural and environmental effects on evacuation time as well. For example, in [20], it was found that the optimal evacuation time needs a combination of herding behaviour and the use of environmental knowledge (about the location of exits). In [47] it was found that when exits are placed symmetrically in a room, the evacuation time is shortest. It was also found that including social elements in the model (finding your group member before exiting, exiting through the entrance, and not wanting to stop but keep moving towards the exit) can make a more robust and realistic model. In [44] the social force model (Helbing social force) was implemented in a cellular automata model to simulate evacuation from a room with one exit. Arching, clogging, and the faster-is-slower-effects were found, showing that the three social forces (repulsion, friction, and attraction) can be basic reasons for complex behaviours emerging from evacuations. Also, changing the width of the door can have a large effect on evacuation time. In [11] it is shown that the crowd density around a person has an impact on that person's speed and that this is an exponential relationship, with more surrounding people reducing the person's speed. In [22] it is shown that evacuation time is not only based on the distance from the exit but also on effects such as the crowd density around the people evacuating and exit choice behaviour. In [27, 28] the social force model was applied. It was found that the wider the doors, the less faster-is-slower-effect there is, because there will be less congestion at the door. Also, the repulsive and dissipative forces seem to have the largest effects on the faster-is-slower-effect. In [19] a lattice gas model of people escaping a smoke filled room was created to replicate the findings of an experiment in which blindfolded students had to find the exit. It was found that adding exits did not shorten evacuation time, but that the evacuation process was based on herding behaviour (following the acoustics). Based on these findings from others, we expect the evacuation time to increase as crowd density increases in our model.

Basic settings simulation experiments. Simulation experiments with different factors and levels were designed to answer different research questions introduced in the following sections. The agent environment chosen for the simulations was a square (20 × 20 m) layout of a building with four exits (top, down, left, right; main exit = down). All environmental and personal factors such as width of the doors, gender, age, and level of compliance were kept constant across simulations. Only the factors and levels stated in each experimental setup in the following sections were systematically varied. The settings that were kept similar, except the few parameters that are structurally changed to answer the current research question, are shown in Table 3 below.

Table 3. Basic parameter settings for the simulation experiments.

Parameter	Setting
Familiarity	50% (i.e. 50% of passengers are familiar with the environment)
Helping	Off
Falls	On
Contagion model	On
Percentage children	15 (based on [29])
Percentage elderly	15 (based on [29])
Percentage people travelling alone	50
Group ratios	33-33-34 (we assume an equal distribution for group sizes)
Percentage females	50%
Fire location	Random location, but always 3 m away from an exit and present from the 1st second
Cultural cluster distribution	Equal division of all passengers over all 11 clusters (9.09% of passengers per cluster)
Length of fall (before standing up)	30 s
Start fire alarm	180 s after the fire starts
Start public announcement	20 s after the fire alarm starts

3.3   Simulation Results: Effect of Falls

Table 4 shows the design of the simulation experiment to determine the effect of falling on evacuation time, total falls, and the average response time. The total number of simulation runs is based on the number of factors and levels, and number of repetitions per combination of factor and level, resulting in $3 \times 2 \times 60 = 360$ simulation runs here. The hypotheses were: (1) when falling behaviour is 'on', evacuation time will be slower than when there are no falls (because it will take extra time to fall and stand back up); (2) when falling behaviour is 'on', falls will happen, but no falls will happen when this feature is turned 'off'; (3) there will be no difference in response times for falling 'on' versus 'off' (as response time precedes evacuation movement).

Evacuation time. The results are shown in Fig. 6. As expected, the higher the crowd density, the slower the evacuation time. Unexpectedly, though, the evacuation time

Table 4. Factors and levels in the simulation experiment for falls.

	Factor	
	Crowd Density	Falls
Level 1	Low	On
Level 2	Medium	Off
Level 3	High	

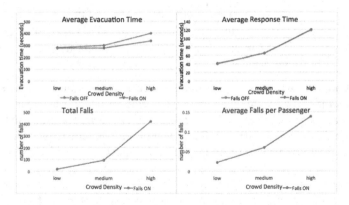

Fig. 6. Effect of falls on evacuation time, falls, and response time.

decreases when falls occur, compared to no falls (see Fig. 6, top left), which is the opposite of what was expected. However, this can be explained due to the fact that the evacuation of the fallen agents is delayed, thereby reducing the overall crowd congestion at exits. Essentially, then, a more phased evacuation takes place, which takes less time. In other words, this could be explained by the faster-is-slower-effect [17]. This effect reflects the observation that certain processes (in evacuation situations, production, traffic dynamics, or logistics) take more time if performed at high speed. In other words, waiting can often help to coordinate the activities of several competing units and thus speed up the overall progress. In our case, falling seems to have similar effects to waiting and speeds up the overall evacuation.

To find out if these effects could be significant, statistical analyses were performed on the data. A 2×3 independent ANOVA was performed on the evacuation time with Falls (with or without) and Crowd Density (low, medium, and high) as between factors. The main effect of Crowd Density was significant, $F(2, 354) = 12.96, p < .001$, and the main effect of Falls was approaching significance, $F(1, 354) = 3.72, p = .055$, but the interaction effect of Falls $\times$ Crowd Density was not significant, $F(2, 354) = 1.23, n.s.$ Post hoc tests with Tukey HSD corrections showed that only high Crowd Density differs significantly from low and medium Crowd Density, but low and medium Crowd Density do not differ significantly: high-low, $p < .001$; high-medium, $p < .001$; low-medium, $n.s.$ In conclusion, then, evacuation time seems to significantly increase for high crowd density versus low or medium crowd density, and a trend is visible for slower evacuation time without falls versus with falls.

Total number of falls. As expected, both the total falls and average falls per person increase as the crowd density increases, for two reasons. First, the more agents there are in the environment, the less room there is to move and so more falling occurs. Second, the more agents there are in the environment, the higher the chances of individuals falling which will increase the average rate (see Fig. 6, bottom row). A 2×3 independent ANOVA was performed on the Total Falls with Falls (with or without) and Crowd Density (low, medium, and high) as between factors. The main effects of Falls and Crowd Density and the interaction effect of Falls $\times$ Crowd Density were significant: $F(2, 354) = 5612.60$, $p < .001$; $F(1, 354) = 11306.25$, $p < .001$; $F(2, 354) = 5612.60$, $p < .001$, respectively. Post hoc tests with Tukey HSD corrections showed that each level of Crowd Density differs significantly from each other level: high-low, $p < .001$; high-medium, $p < .001$; low-medium, $p < .001$.

Response time. As expected, response time increases as crowd density increases and no significant differences were found in response time for falling behaviour 'on' versus 'off'. Statistical analyses confirm these findings. A 2×3 independent ANOVA was performed on the response time with Falls (with or without) and Crowd Density (low, medium, and high) as between factors. The main effect of Crowd Density was significant, $F(2, 354) = 4773.30$, $p < .001$. There was no main effect of Falls, $F(1, 354) = .012$, n.s., and no interaction effect of Falls $\times$ Crowd Density, $F(2,354) = .681$, n.s. Post hoc tests with Tukey HSD corrections show that each level of Crowd Density differs significantly from the other two: low-medium, $p < .001$; medium-high, $p < .001$; low-high, $p < .001$.

3.4   Simulation Results: Helping Behaviour

Table 5 shows the design of the simulation experiment to determine the effect of helping behaviour on evacuation time, falls, and response time, resulting in $3 \times 2 \times 60 = 360$ simulation runs here. The hypotheses were: (1) when people help others, the evacuation time is longer than when people do not help (because the helpers will take more time to evacuate; although only a small effect is expected); (2) when passengers help others, the number of falls will increase (because the helpers next to the fallen passengers create more obstacles; although only a small effect is expected); (3) no difference is expected in response times for helping 'on' versus 'off' (because the decision to evacuate precedes helping).

Table 5. Factors and levels in the simulation experiment for crowd density and helping.

	Factor	
	Crowd Density	Helping
Level 1	Low	On
Level 2	Medium	Off
Level 3	High	

Evacuation time. The results are shown in Fig. 7. As expected, evacuation time increases as crowd density increases. However, unexpectedly, helping behaviour seems to reduce evacuation time for high crowd density environments slightly, but not for low to medium crowd density. This could be explained by those helping delaying their evacuation slightly and forming less congestion overall, like a phased evacuation, as happened with the falls. Essentially, people will evacuate one after another (sequentially) which creates less congestion at the doors (see Fig. 7, left). Again, this could be explained with the faster-is-slower-effect, mentioned in the explanation of falls, reducing the average evacuation time [17]. When analysing these effects statistically,

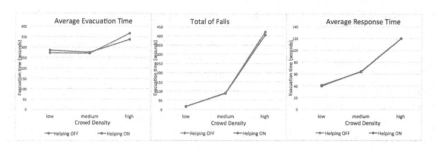

Fig. 7. Effect of helping behaviour on evacuation time, falls, and response time.

though, only the main effect of crowd density is significant and not the effect of helping. A 2 × 3 independent ANOVA was performed on the response time with Helping (with or without) and Crowd Density (low, medium, and high) as between factors. The main effect of Crowd Density was significant, $F(2, 354) = 22.87$, $p < .001$. However, there was no main effect of helping, $F(1, 354) = .119$, *n.s.*, and no interaction effect of Falls × Crowd Density, $F(2, 354) = 1.37$, *n.s.* Post hoc tests with Tukey HSD corrections show that only high Crowd Density differs significantly from low and medium Crowd Density, and low and medium Crowd Density do not differ significantly: high-low, $p < .001$; high-medium, $p < .001$; low-medium, *n.s.*

Total number of falls. The number of falls naturally increases as the crowd density increases. This increase seems similar for helping behaviour 'on' and 'off', but the difference is actually significant when tested statistically (see Fig. 7, middle). A 2 × 3 independent ANOVA was performed on the total Falls with Helping (with or without) and Crowd Density (low, medium, and high) as between factors. The main effects of Crowd Density, $F(2, 354) = 22.87$, $p < .001$, and Helping were significant, $F(1, 354) = 8.45$, $p < .01$, as was the interaction effect of Helping × Crowd Density, $F(2, 354) = 5.52$, $p < .01$. Post hoc tests with Tukey HSD corrections show each level of Crowd Density differs significantly from each other: low-medium, $p < .001$; medium-high, $p < .001$; low-high, $p < .001$. In conclusion, the number of falls increases both when crowd density increases and also without helping.

Response time. As expected, no differences are observed in the average response times for helping behaviour 'on' and 'off', only an effect of crowd density which statistical analyses confirm. A 2 × 3 independent ANOVA was performed on the Response Time with Helping (with or without) and Crowd Density (low, medium and high) as between factors. The main effect of Crowd Density was significant, $F(2, 354) = 5162.73$, $p < .001$, while neither the main effect of Helping, $F(1, 354) = .416$, *n.s.*, or the interaction effect of Helping × Crowd Density were significant, $F(2, 354) = .798$, *n.s.* Post hoc tests with Tukey HSD corrections show each level of Crowd Density differs significantly from each other: low-medium, $p < .001$; medium-high, $p < .001$; low-high, $p < .001$ (see Fig. 7, right).

3.5   Experimental Results: Social Contagion and Familiarity

Table 6 shows the experimental design of the simulation experiment to determine the effect of social contagion and familiarity on evacuation time, falls, and response time, resulting in $3 \times 3 \times 2 \times 60 = 1080$ simulation runs here. The hypotheses were: (1) evacuation time will be faster with social contagion than without (because people will still find out from others there is a fire, even when not observed personally); (2) when crowd density increases, there will be more falls; (3) when there is social contagion, there will be fewer falls (because without it, more people will find out the situation is dangerous through the fire alarm, which means more people will evacuate simultaneously, thereby falling more); (4) response time will be faster with social contagion than without (because people who do not observe the fire themselves are informed faster by others); (5) response time will be faster the more familiar people are with the environment (because taking the nearest exit in combination with social contagion will speed up the response time, spreading the 'news' faster than when people all take the same exit); and finally (6) the higher the crowd density, the slower the response time.

Table 6. Factors and levels in the simulation experiment for social contagion and familiarity

	Factor		
	Crowd Density	Familiarity	Social Contagion
Level 1	Low	0%	On
Level 2	Medium	50%	Off
Level 3	High	100%	

Evacuation time. The results are shown in Fig. 8. As expected, with social contagion there is a decrease in evacuation time compared to without, and the more familiar people are with the environment, the faster their evacuation time (see Fig. 8, top row), which statistical analyses confirmed. The social contagion of mental and emotional states is a form of collective group decision making or collective intelligence [17]. It is also related to herding, as individuals are 'infected' with other's decisions and follow them when their own intentions are not as strong as those of others around them. [16]. A 2×3 independent ANOVA was performed on Evacuation Time with Social Contagion (with or without) and Crowd Density (low, medium, and high) as between factors. The main effects of Crowd Density and Social Contagion and the interaction effect of Social Contagion $\times$ Crowd Density were significant: $F(2, 354) = 133.81$, $p < .001$; $F(1, 354) = 237.76$, $p < .001$; $F(2, 354) = 4.35$, $p < .05$, respectively. Post hoc tests with Tukey HSD corrections show each level of Crowd Density differs significantly from each other: low-medium, $p < .05$; medium-high, $p < .001$; low-high, $p < .001$. A 3×3 independent ANOVA was performed on the Evacuation Time with Familiarity (0%, 50%, or 100%) and Crowd Density (low, medium, and high) as between factors. The main effects of Crowd Density and Familiarity and the interaction effect of Familiarity $\times$ Crowd Density were significant: $F(2, 354) = 125.83$; $p < .001$, $F(1, 354) = 23.16$, $p < .001$; $F(2, 354) = 31.10$, p < .001, respectively. Post hoc tests

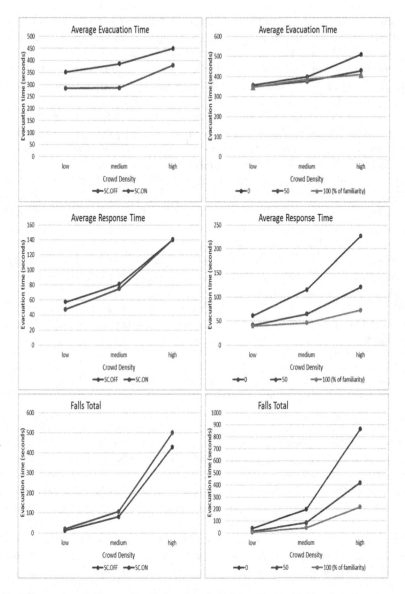

Fig. 8. Effects of social contagion (left column) and familiarity (right column) on evacuation time, response time, and falls.

with Tukey HSD corrections show each level of Crowd Density differs significantly from each other: low-medium, $p < .05$; medium-high, $p < .001$; low-high, $p < .001$. For Familiarity, only 0% familiarity differs significantly from 50% and 100%, but not 50% from 100%: 0%–50% $p < .05$; 50%–100% $n.s.$; 0%–100% $p < .05$.

Total number of falls. As expected, the number of falls is lower with social contagion than without. This can be explained by people starting to evacuate earlier, spreading the evacuation across the simulation time. Consequently, there are fewer collisions among passengers, which result in fewer falls. Familiarity shows the same effect: the more familiar the crowd members are with the environment, the more distributed among the exits they are, which consequently leads to fewer collisions and falls (see Fig. 8, bottom row). Statistical analyses confirmed these interpretations. A 2 × 3 independent ANOVA was performed on the Total Falls with Social Contagion (with or without) and Crowd Density (low, medium, and high) as between factors. The main effects of Crowd Density and Social Contagion and the interaction effect of Social Contagion × Crowd Density were significant: $F(2, 354) = 732.98$, $p < .001$; $F(1, 354) = 11.88$, $p < .01$; $F(2, 354) = 3.42$, $p < .05$. Post hoc tests with Tukey HSD corrections show each level of Crowd Density differs significantly from each other: low-medium, $p < .001$; medium-high, $p < .001$; low-high, $p < .001$. A 3 × 3 independent ANOVA was performed on the Total Falls with Familiarity (0%, 50%, or 100%) and Crowd Density (low, medium, and high) as between factors. The main effects of Crowd Density and Familiarity and the interaction effect of Familiarity × Crowd Density were significant: $F(2, 354) = 17290.13$; $p < .001$; $F(1, 354) = 6227.45$, $p < .001$; $F(2, 354) = 3062.52$, $p < .001$. Post hoc tests with Tukey HSD corrections show each level of Crowd Density and Familiarity differs significantly from each other: low-medium, $p < .001$; medium-high, $p < .001$; low-high, $p < .001$; 0%–50%, $p < .001$; 50%–100%, $p < .001$; 0%–100%, $p < .001$.

Response time. As expected, response time increases as crowd density increases and with social contagion the increase is lower than without social contagion. Similarly, the more familiar people are with their environment, the less the response time increases as crowd density increases. This is explained by familiarity distributing people over the available exits, which helps to convey the fear and belief of danger with social contagion to others who start to evacuate early (see Fig. 8, middle row). Statistical analyses confirmed the two main effects of crowd density and social contagion. A 2 × 3 independent ANOVA was performed on Response Time with Social Contagion (with or without) and Crowd Density (low, medium, and high) as between factors. The main effects of Crowd Density, $F(2, 354) = 410.46$, $p < .001$, and Social Contagion were significant, $F(1, 354) = 4.46$, $p < .05$, while the interaction effect of Social Contagion × Crowd Density was not significant, $F(2, 354) = 1.16$, $n.s.$ Post hoc tests with Tukey HSD corrections show each level of Crowd Density differs significantly from each other: low-medium, $p < .001$; medium-high, $p < .001$; low-high, $p < .001$. A 3 × 3 independent ANOVA was performed on Response Time with Familiarity (0%, 50%, or 100%) and Crowd Density (low, medium, and high) as between factors. The main effects of Crowd Density and Familiarity and the interaction effect of Familiarity × Crowd Density were significant: $F(2, 354) = 11785.94$, $p < .001$; $F(1, 354) = 10311.63$, $p < .001$; $F(2, 354) = 2334.88$, $p < .001$, respectively. Post hoc tests with Tukey HSD corrections show each level of Crowd Density and Familiarity differs significantly from each other: low-medium, $p < .001$; medium-high, $p < .001$; low-high, $p < .001$; 0%–50%, $p < .001$; 50%–100%, $p < .001$; 0%–100%, $p < .001$.

3.6   Groups

Table 7 shows the design of the simulation experiment to determine the effect of group size on evacuation time, falls, and response time, resulting in $3 \times 4 \times 60 = 720$ simulation runs. The hypotheses were: (1) the more people who travel alone, the faster the evacuation time (because people will move faster by themselves); (2) the bigger the groups, the slower the evacuation time (although this is expected to be a small effect); (3) the more people who travel alone, the fewer falls (because groups form more congestion; although this is expected to be a small effect); (4) the larger the groups, the more falls (because of more congestion); (5) the more people who travel alone, the faster the response time (because people can evacuate faster); and (6) the bigger the groups, the slower the response time (although this is expected to be a small effect).

Table 7. Factors and levels in the simulation experiment for groups

	Factor	
	Crowd Density	Travelling Alone
Level 1	Low	100%
Level 2	Medium	0% (only groups of 2 adults)
Level 3	High	0% (only groups of 3 adults)
Level 4		0% (only groups of 4 adults)

Evacuation time. The results are shown in Figs. 9 and 10. As expected, as crowd density increases, evacuation time becomes slower. Unexpectedly, though, it seems that people travelling alone and in groups of three are slower to evacuate than groups of two and four. Indeed, groups of four evacuate the fastest and people travelling alone are actually slowest (Fig. 9). Statistical analysis confirms this interpretation. A 4×3 independent ANOVA was performed on Evacuation Time with Group Size (1, 2, 3, and 4) and Crowd Density (low, medium, and high) as between factors. The main effects of Crowd Density and Group Size, and the interaction effect of Group Size $\times$ Crowd Density were significant: $F(2, 354) = 22643.44, p < .001; F(3, 354) = 137.15, p < .001; F(6, 354) = 3.70, p < .001$. Post hoc tests with Tukey HSD corrections show

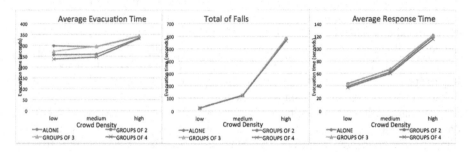

Fig. 9. Effects of groups on evacuation time, falls, and response time.

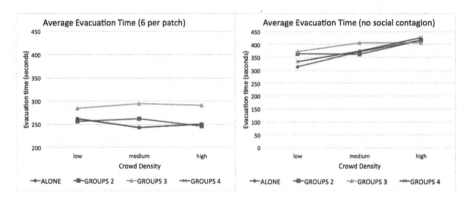

Fig. 10. Effects of groups on evacuation time with a maximum travel capacity of six people per m² (left) and without social contagion (right).

that only high Crowd Density differs significantly from low and medium Crowd Density: high-low, $p < .001$; high-medium, $p < .001$; low-medium, n.s. For Group Size, these tests show that a lone person does not differ from groups of 3, and groups of 2 do not differ from groups of 4; however, all others differ significantly from each other: 1–2, $p < .001$; 1–3, n.s.; 1–4, $p < .001$; 2–3, $p < .01$; 2–4, n.s. In conclusion, evacuation time increases when crowd density increases and decreases for groups of 4 and 2 versus groups of 3 or 1.

This is unexpected and seems to not be an effect of speed, because all group sizes have the same number of falls. Therefore, it does not seem to be a faster-is-slower-effect [17]. When inspecting the average speed during simulations, it was confirmed that they did not differ for group sizes. Also, the outcome measures did not differ significantly for different numbers of children and elderly, which could influence the average speeds of the groups. However, what could explain groups of four being faster than people travelling alone is social contagion in combination with moving through space. With social contagion, or collective intelligence, groups can 'infect' each other faster with emotions and beliefs, compared to people travelling alone, which is beneficial for evacuation time. Moving through space is implemented with a maximum of 8 passengers per patch (square metre), meaning lone passengers and groups of 2 and 4 can always use a patch to its maximum capacity, but groups of 3 can only fit a maximum of two groups (6 passengers) per patch at one time step. This means groups of 3 are a little disadvantaged, since groups of 1, 2, and 4 can always move around in space with maximum capacity. That could explain why groups of three and people alone are slowest and groups of 2 and 4 are fastest. We have tested this by running similar simulation experiments like this one, but then (1) without social contagion, and (2) with a maximum capacity of 6 people per square metre. The expectation is that (1) without contagion, groups of 3 will be slowest versus groups of 1, 2 and 4, and (2) with a maximum capacity of 6 people per square metre, groups of 4 will be slowest compared to people travelling alone and groups of 2 and 3. As expected, without social contagion, groups of 3 are slowest in evacuation time (see Fig. 10). No effects of falls and response time were observed in this experiment. Unexpectedly, groups of 3 are not the fastest with a maximum capacity of 6 per square metre, but again the slowest. This means that social contagion is only part of the explanation for groups being slower to evacuate than people travelling alone. We cannot find more explanations for this in the literature because (1) the impact of groups on crowd dynamics is still largely unknown [24, 31], and (2) we have not modelled group formations, such as in [24], that could influence the crowd dynamics. We have chosen to model a group as moving through space as a 'square' group, with all members moving from patch (square metre) to patch simultaneously. So, group formations are no explanation either. However, social contagion is part of the effect of groups of 2 and 4 being faster than people travelling alone or in groups of 3.

Total number of falls. As crowd density increases, the number of falls increase; although no significant differences were found between group sizes, as expected. Statistical analysis confirmed this interpretation of the graph. A 4×3 independent ANOVA was performed on Total Falls with Group Size (1, 2, 3, and 4) and Crowd Density (low, medium, and high) as between factors. The main effect of Crowd Density was significant, $F(2, 354) = 24048.28$, $p < .001$, but the main effect of Group Size, $F(3, 354) = 1.39$, $n.s.$, and the interaction effect of Group Size $\times$ Crowd Density were not significant, $F(6, 354) = 1.93$, $n.s.$ Post hoc tests with Tukey HSD corrections show that each level of Crowd Density differs significantly from each other: low-medium, $p < .001$; medium-high, $p < .001$; low-high, $p < .001$.

Response time. As crowd density increases, response time increases. Although no significant differences between group sizes were expected, statistical analysis showed that groups of 2 and 4 are faster in their response time than groups of 1 and 3. This seems plausible as it is similar with the evacuation time, which both can be explained by the social contagion effects. A 4×3 independent ANOVA was performed on Response Time with Group Size (1, 2, 3, and 4) and Crowd Density (low, medium, and high) as between factors. The main effects of Crowd Density, $F(2, 354) = 9634.55$, $p < .001$, and Group Size were significant, $F(3, 354) = 43.73$, $p < .001$, and the interaction effect of Group Size $\times$ Crowd Density was not, $F(6, 354) = .467$, $n.s.$ Post hoc tests with Tukey HSD corrections show that each level of Crowd Density differs significantly from each other: low-medium, $p < .001$; medium-high, $p < .001$; low-high, $p < .001$; and group size 1 and 3 do not differ significantly, while the other group sizes do: 1–2, $p < .001$; 1–3, $n.s.$; 1–4, $p < .001$; 2–3, $p < .001$; 2–4, $p < .001$; 3–4, $p < .001$. Taking all these results into account, it seems that social contagion is the biggest cause for the group effects.

3.7   Age

Table 8 shows the design of the simulation experiment, resulting in $3 \times 2 \times 60 = 360$ simulation runs here. The hypotheses were: (1) elderly people have slower evacuation times, compared to adults (because elderly people move slower); (2) there will be no differences in number of falls between adults and elderly people; (3) there will be no differences in response time between adults and elderly people.

Table 8. Factors and levels in the simulation experiment for age

	Factor	Age
	Crowd Density	Age
Level 1	Low	Travelling alone 100% adults
Level 2	Medium	Travelling alone 100% elderly
Level 3	High	

Evacuation time. The results are shown in Fig. 11. As crowd density increases, so does evacuation time. As expected, elderly people seem to be slower in evacuating than adults, most likely due to their slower movement. In this experiment, all passengers are elderly or adults exclusively, so the exact same effects are there with the elderly as with adults. For instance, there is no faster-is-slower-effect [17] here for age, because that would require differences in speed within the same simulation run. So, in this case, faster speed does mean faster evacuation. Here, the faster-is-slower-effect was present for the adults by themselves, but as a result of falls, again. However, the elderly did not fall based on their slower speeds, which in turn prevented a faster-is-slower-effect for them based on falls (see Fig. 11). Indeed, statistical analysis showed there was an effect of age. A 2 × 3 independent ANOVA was performed on Evacuation Time with Age (adult, elder) and Crowd Density (low, medium, and high) as between factors. Both the main effects of Crowd Density, $F(2, 354) = 35.40, p < .001$, and Age were significant, $F(1, 354) = 3.20, p < .001$, but the interaction effect of Age × Crowd Density was not significant, $F(2, 354) = .359, n.s.$ Post hoc tests with Tukey HSD corrections show that each level of Crowd Density differs significantly from each other level: low-medium, $p < .001$; medium-high, $p < .001$; low-high, $p < .001$.

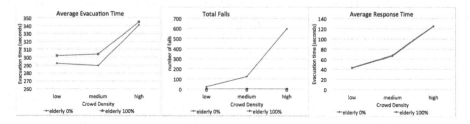

Fig. 11. Effects of age (speed) on evacuation time, falls, and response time

Total number of falls. As expected, as crowd density increases, the number of falls increases. Unexpectedly and very interestingly, elderly people have no falls and the falls of the adults increase as crowd density increases. No falls for elderly people seems unrealistic in real life, however, because elderly people should be more prone to falling than adults. The explanation for this finding is based on how falls are implemented in this model. Currently, they are based on the speed of the passengers and their age is not taken into account, so this could be improved in a future version on the IMPACT model. Discounting age, based on speed alone it makes sense that passengers who move slower have fewer falls (see Fig. 11). Statistical analysis confirmed these interpretations of the graphs. A 2 × 3 independent ANOVA was performed on Total Falls with Age (adult, elder) and Crowd Density (low, medium, and high) as between factors. The main effects of Crowd Density, $F(2, 354) = 13245.73$, $p < .001$, and Age were significant, $F(1, 354) = 26056.94$, $p < .001$, and the interaction effect of Age × Crowd Density was also significant, $F(2, 354) = 13245.73$, $p < .001$. Post hoc tests with Tukey HSD corrections show that each level of Crowd Density differs significantly from each other level: low-medium, $p < .001$; medium-high, $p < .001$; low-high, $p < .001$.

Response time. As expected, as crowd density increases, response time becomes slower. Also, as expected, the response time does not differ significantly between the elderly and adults (see Fig. 11). Statistical analysis confirmed this interpretation of the graph. A 2 × 3 independent ANOVA was performed on Response Time with Age (adult, elder) and Crowd Density (low, medium, and high) as between factors. The main effect of Crowd Density was significant, $F(2, 354) = 5507.43$, $p < .001$; however, the main effect of Age, $F(1, 354) = 2.52$, *n.s.*, and the interaction effect of Age × Crowd Density were not significant, $F(2,354) = .03$, *n.s.* Post hoc tests with Tukey HSD corrections show that each level of Crowd Density differs significantly from each other level: low-medium, $p < .001$; medium-high, $p < .001$; low-high, $p < .001$.

3.8   Compliance

Table 9 shows the design of the simulation experiment to determine the effect of compliance on evacuation time, number of falls, and response time, resulting in $3 \times 2 \times 60 = 360$ simulation runs here. The hypotheses were: (1) evacuation time is faster for 100% compliance than 0% compliance; (2) more falls will happen with 100% compliance compared with 0% (because people will evacuate faster resulting in crowding and so more falls); (3) response time will be faster for 100% compliance compared to 0% (because people will decide to evacuate faster). This simulation experiment was also run for adults and the elderly, both female and male. With the current parameter settings, no significant differences between females and males or adults and the elderly were found, meaning that the difference in the current compliance level settings for gender and age do not create differences in the actions (see Sect. 2.1 for these settings). Therefore, to find the effect of the compliance parameter, this experiment was set up comparing a low with a high level. For a maximum effect of compliance, levels 1 and 0 were preferred, but the simulation does not run with

Table 9. Factors and Levels in the Simulation Experiment for Compliance

	Factor	
	Crowd Density	Compliance
Level 1	Low	Compliance level 0.1 (only male adults)
Level 2	Medium	Compliance level 1 (only male adults)
Level 3	High	

compliance set to 0, since the passengers will not move then. Compliance set to 0.001 or 0.01 resulted in one simulation run taking multiple days. With the value of 0.1 there is still a large effect of compliance to be seen and the simulation runs were practically feasible to run, so this level was selected for the experiment.

Evacuation time. Results are shown in Fig. 12. As expected, as crowd density increases, evacuation time increases, and high compliance results in faster evacuation time than low compliance. A 2 × 3 independent ANOVA was performed on Evacuation Time with Compliance (low, high) and Crowd Density (low, medium, and high) as between factors. The main effects of Crowd Density and Compliance and the interaction effect of Compliance × Crowd Density were all significant: $F(2, 354)$ = 33.75, $p < .001$; $F(1, 354) = 3092.49$, $p < .001$; $F(2,354) = 6.65$, $p < .001$, respectively. Post hoc tests with Tukey HSD corrections show that each level of Crowd Density differs significantly from each other level: low-medium, $p < .01$; medium-high, $p < .001$; low-high, $p < .001$.

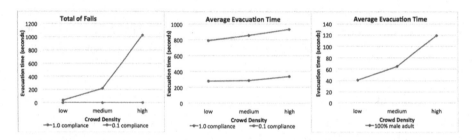

Fig. 12. Effects of compliance on evacuation time, falls, and response time.

Total number of falls. As expected, more falls happen as crowd density increases and when there is high compliance versus low compliance. No falls happened in the low compliance simulation runs, though, which can be explained by the slower speed that is a result of low compliance. A 2 × 3 independent ANOVA was performed on Total Falls with Compliance (low, high) and Crowd Density (low, medium, and high) as between factors. The main effects of Crowd Density and Compliance and the interaction effect of Compliance × Crowd Density were all significant: $F(2, 354) = 13110.60$, $p < .001$; $F(1, 354) = 25825.15$, $p < .001$; $F(2,354) = 13110.60$, $p < .001$, respectively. Post hoc tests with Tukey HSD corrections show that each level of Crowd Density differs significantly from each other level: low-medium, $p < .001$; medium-high, $p < .001$; low-high, $p < .001$.

Response time. Response times for male adults are shown in Fig. 12, which do not significantly differ from female adults and the elderly, as expected, and show a similar pattern for a high compliance level. The response time for the low compliance level did not register in the simulations; that is why the response time for male adults with a compliance level of 0.89 are shown and analysed. An independent one-way ANOVA was performed on the Response Time of male adults with Crowd Density (low, medium, and high) as a between factor. The main effect of Crowd Density was significant, $F(2, 717) = 397678.37$, $p < .001$. Post hoc tests with Tukey HSD corrections show that each level of Crowd Density differs significantly from each other level: low-medium, $p < .001$; medium-high, $p < .001$; low-high, $p < .001$.

3.9   Environment

Table 10 shows the design of the simulation experiment to determine the effect of room type on evacuation time, falls, and response time, resulting in $3 \times 6 \times 60 = 1080$ simulation runs here. The hypotheses were: (1) evacuation time increases faster in the rectangular room than the square room (because people take more time to reach the exits); (2) the number of falls is higher in the rectangular room (because people use more steps to reach the exits); (3) response time is slowest in the rectangular room (because in larger rooms there is less chance of observing the fire).

Table 10. Factors and Levels in the Simulation Experiment for Environment

	Factor	
	Crowd Density	Room type
Level 1	Low	Type 1 (square, 20 × 20 m)
Level 2	Medium	Type 2 (rectangle 20 × 40 m)
Level 3	High	

Evacuation time. Results are shown in Fig. 13. As expected, evacuation time increases as crowd density increases, although this only happened for high crowd density and not low and medium densities (see Fig. 13, left). Statistical tests confirm this interpretation of the graph. A 2×3 independent ANOVA was performed on Evacuation Time with Room Type (square or rectangle) and Crowd Density (low, medium, and high) as between factors. The main effects of Crowd Density and Room Type and the interaction effect of Room Type × Crowd Density were all significant: $F(2, 354) = 104.97$, $p < .001$; $F(1, 354) = 443.17$, $p < .001$; $F(2,354) = 35.07$, $p < .001$, respectively. Post hoc tests with Tukey HSD corrections show that high

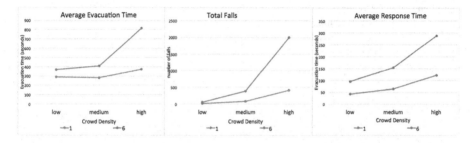

Fig. 13. Effects of room type on evacuation time, falls, and response time.

Crowd Density differs significantly from low and medium, but low and medium do not differ significantly: low-medium, *n.s.*; medium-high, $p < .001$; low-high, $p < .001$.

Total number of falls. As crowd density increases, so do the number of falls. The number of falls also increase faster in the larger room than in the smaller room.

Table 11. Effects of Socio-Cultural, Cognitive, and Emotional Elements on Evacuation Time

Model element	Variations	Average evacuation time (seconds)	Difference from benchmark (seconds)	Relative difference from benchmark (percentage)
Falls	Off (benchmark)	324.31		
	On	293.51	−30.8	−9.5%
Helping behaviour	Off (benchmark)	302.57		
	On	298.86	−3.71	−1.2%
Social Contagion	Off (benchmark)	396.12		
	On	317.27	−78.85	−20.0%**
Familiarity	0% of passengers familiar with environment (benchmark)	412.47		
	50% of passengers familiar with environment	385.38	−27.09	−6.6%***
	100% of passengers familiar with environment	381.52	−30.95	−7.5%***
Groups	People travelling alone (benchmark)	311.29		
	Groups of two	282.95	−28.37	−9.1%***
	Groups of three	303.87	−7.42	−2.4%***
	Groups of four	217.79	−93.5	−30.0%***
Age	All adults (benchmark)	307.6		
	All elderly people	316.83	+9.23	+3.0%***
Compliance	High compliance (1.0) (benchmark)	301.17		
	Low compliance (0.1)	856.03	+554.86	+184.2%***
Environment	Small square room (20 × 20 m) (benchmark)	313.07		
	Big rectangle room (20 × 40 m)	530.86	+217.79	+705.0%***

Significant main effect: **$p < .01$, ***$p < .001$.

Note that the increase in falls is not due to more space in the rectangular room and more space to move (faster) towards the exits, as the crowd densities are kept the same relative to the total square metres of the room. Rather, a longer pathway (more steps towards the exit) increases the chance of falling (see Fig. 13, middle). Statistical analysis confirms this interpretation of the graph. A 2×3 independent ANOVA was performed on Total Falls with Room Type (square or rectangle) and Crowd Density (low, medium, and high) as between factors. The main effects of Crowd Density and Room Type and the interaction effect of Room Type $\times$ Crowd Density were all significant: $F(2, 354) = 2100.66$, $p < .001$; $F(1, 354) = 1524.03$, $p < .001$; $F(2, 354) = 893.53$, $p < .001$. Post hoc tests with Tukey HSD corrections show that each level of Crowd Density differs significantly from each other level: low-medium, $p < .001$; medium-high, $p < .001$; low-high, $p < .001$.

Response time. As expected, the response time is slower in the rectangular room than in the square room and also increases as crowd density increases (see Fig. 13, right). Statistical analysis confirms this interpretation of the graph. A 2×3 independent ANOVA was performed on Response Time with Room Type (square or rectangle) and Crowd Density (low, medium, and high) as between factors. The main effects of Crowd Density and Room Type and the interaction effect of Room Type $\times$ Crowd Density were all significant: $F(2, 354) = 5648.72$, $p < .001$; $F(1, 354) = 11279.66$, $p < .001$; $F(2, 354) = 1003.42$, $p < .001$, respectively. Post hoc tests with Tukey HSD corrections show that each level of Crowd Density differs significantly from each other level: low-medium, $p < .001$; medium-high, $p < .001$; low-high, $p < .001$.

3.10   Comparing Results: Influence of Socio-Cultural, Cognitive, and Emotional Elements

In this section, the effects of the socio-cultural, cognitive, and emotional elements in the model will be compared to identify how much each element influences the total evacuation time. In this way, the added value of each element can be interpreted. Of course, this is in the case of the empty environment studied in the simulation experiments, where only the human behaviour is studied during evacuation. In real life, the effects of the socio-cultural, cognitive, and emotional elements will be combined with environmental influences, such as obstacles, stairs, corridors, lanes, and pathways. Table 11, above, shows the effects of each model element (e.g. falling, helping, social contagion) on the total evacuation time in seconds and is expressed as a percentage of relative difference compared to the benchmark. The relative differences of each model element range from reducing the total evacuation time by 30% to increasing it by 705%. Most notable are the decreases in evacuation time caused by social contagion of 20%, familiarity of between 6.6% and 7.5%, and travelling in groups of between 2.4% and 30%. Compliance and environment type also have a very large effect on the evacuation time – increasing it by 184.2% and 705%, respectively – but these two effects are harder to compare in size with the others in the table, because the parameter settings of compliance and the sizes of the environment types made the effect very large. The other effects are comparable, though, because the human behaviour all takes place in the same environment and the settings chosen are realistic. In conclusion, the socio-cultural, cognitive, and emotional elements that can be compared – falling, helping, social contagion, familiarity with environment, group sizes, and age – have an effect on evacuation time between decreasing it by 30% to increasing it by 3%.

4   Conclusion and Discussion

The aim of this research was to create and validate an evacuation simulation that includes socio-cultural, cognitive, and emotional factors in response to the need for more realistic crowd models that incorporate psychological and social factors. The development of the model drew on insights from social and cross-cultural psychology, interviews with crisis management experts, and was based on scientific findings and literature. The model was validated against data from an evacuation drill simulated by the existing EXODUS evacuation model [13, 26]. Our IMPACT model was compared with this benchmark on multiple outcome measures and results showed that, on all measures, the IMPACT model was within or close to the prescribed boundaries, thereby establishing its validity.

Next, multiple simulation experiments were run to answer research questions concerning the effects of the socio-cultural, cognitive, and emotional elements in the model on evacuation time, total number of falls, and response time. Important findings are that emergent effects, such as the faster-is-slower-effect [17], were found in our results in new forms: as effects of falling, helping, social contagion, and familiarity with the environment. For instance, both falling behaviour and helping (in high crowd density) led to faster evacuation times. The explanation is that falling and helping create a more phased evacuation – as the delays they cause effectively stagger the evacuation and reduce congestion – that results in a faster overall process. Moreover, as expected, social contagion also creates faster evacuation times, because information about the need to evacuate spreads faster than without social contagion. It also unexpectedly led to less falls, which again can be explained by the faster-is-slower-effect. Again, like with falls and helping, people are more phased in their evacuation, meaning less congestion at the bottlenecks (the exits) and therefore less falls. Furthermore, the more people are familiar with the environment: (1) the faster the evacuation time, (2) the fewer the falls, and (3) the faster the response time. These results are a combination of a phased evacuation (meaning less congestion and fewer falls, and therefore a faster-is-slower-effect resulting in faster evacuation time), less congestion (more people spread through the environment going to the nearest exits instead of all taking the same exit, meaning fewer falls), and social contagion (the decision to evacuate can spread faster, meaning faster response times and evacuation times). Groups also showed an interesting effect. The current model suggests it is actually faster to evacuate in groups than alone. This was not based on speed, and therefore not a faster-is-slower-effect, but partly based on social contagion (collective intelligence and herding). The impact of groups on crowd dynamics is still largely unknown [24] and we have not modelled group formations, such as in [24], that could influence the crowd dynamics. Rather, we had chosen to model a group as moving through space as a 'square' group, with all members moving from patch (square metre) to patch simultaneously. The effect of group formations would therefore need further research with the current model.

The faster-is-slower-effect was not found when comparing age groups, however, as the elderly evacuated more slowly than adults although moving more slowly. The reason for the faster-is-slower-effect not being present here for age is that it would require differences in speed within the same simulation run. In this model, however, all passengers within a simulation were either exclusively fast (adults) or slow (elderly people), which meant that faster speed means faster evacuation here. For adults by themselves the faster-is-slower-effect was present, but then as a result of falls. The elderly did not fall due to their slower speeds, which in turn prevented a faster-is-slower-effect when looking at falls instead of speed. The elderly did not fall once in the simulation which is not realistic in real life, since elderly people are more prone to falling. The current implementation of falling is based on speed alone and therefore needs to be improved to also take age into account. With a high level of compliance, people evacuate faster than with a low level of compliance, as expected. The current settings of compliance levels do not make enough differentiation between different ages and genders to have an effect. The simulation experiment showed that the compliance parameter can have an effect, but not with the current model settings. It needs to be decided if this parameter can be omitted or if new parameter settings for different ages and genders can be calculated from new data. Finally, in the smaller square room (20 × 20 m), evacuation was faster than in the larger rectangular room (20 × 40 m). Also, in the smaller square room there were fewer falls and a faster average response time than expected. Essentially, taking more steps towards the exit means more chance of falling.

Comparing all simulation results together, the socio-cultural, cognitive, and emotional elements have an effect from reducing evacuation time by 30% through to increasing it by 3% when the following model elements are considered: falling, helping, social contagion, familiarity with environment, group sizes, and age. However, the parameter settings of compliance and the sizes of the environment types made these effects very large (increasing evacuation time up to 705%) and are therefore left out in this comparison. Overall, this demonstrates that including socio-cultural, cognitive, and emotional elements in evacuation models is both feasible and vital, as they can influence evacuation time by up to 30%. Of course, this is only based on our experiments in an empty square room, where there is no interaction with environmental features such as obstacles, corridors, counterflows, stairs, and others. Therefore, this (maximum) 30% effect on evacuation time should be seen as a 'pure' effect of the socio-cultural, cognitive, and emotional elements in the model, without these additional environmental influences.

The strengths of this research are the inclusion of psychological and socio-cultural aspects in the crowd simulation model, based on research literature and support from stakeholders. Furthermore, the statistical analyses of the experimental results strengthen the interpretations. The current weaknesses of this work are that not every socio-cultural, cognitive, and emotional parameter that was identified during the development of the model is yet implemented to test, such as passengers' disabilities. Conversely, though, the more parameters in the model, the more complex it becomes, and the more difficult it is to analyse and interpret all the results, so there are also benefits to this. Furthermore, the results of the simulations cannot be taken for granted and they will naturally remain estimations. However, because the simulations are based on sufficient background literature, and research and interaction with stakeholders, we believe them to be sound estimations. Moreover, the work limits itself to making predictions about the influence of human behaviour on the evacuation process. All the socio-cultural, emotional, and cognitive effects were tested in an empty room with four exits. In real life, these effects would be combined with the influences of the environment itself, such as corridors, number of exits, stairs, and obstacles. This research could therefore be extended by investigating the combined effect of these elements with the environment, like in [42]. Also, a very important phenomenon – counterflow – was not modelled here. In the current model, all passengers can always take their own pathway towards an exit and do not have to cross or overtake others in the simulation. Therefore, the effects of counterflows are not modelled. Also, it was assumed that when people fall they can stand up again after a while. In reality, people could be trampled on or injure themselves and therefore not be able to stand up again. Consequently, the way we modelled falling behaviour here is just a first step towards studying this effect. However, it is difficult to model, since there is no research conducted yet (to the knowledge of the authors) that indicates what the chances of falling are in certain crowd densities and environments, and also how long it takes to stand back up. Future work consists of developing the model further to simulate realistic transport hub environments and extending the pathfinding behaviour with more heuristics.

To conclude, we reiterate three points that summarise our findings and implications: (1) our model is a realistic evacuation simulation, validated in comparison with an established model and demonstrating well-known emergent effects, such as the faster-is-slower-effect; (2) we would recommend that evacuation simulation modellers include socio-cultural, emotional, and cognitive elements in future models, based on the substantial effect sizes found here (reducing evacuation time by up to 30%), especially social contagion; (3) cultural and social diversity can be beneficial to evacuation as they create more phased evacuations, which create an overall benefit from the faster-is-slower-effect. Further implications are that transport operators, emergency managers, and prevention professionals can use these kinds of agent-based models to predict outcomes and inform decision making when designing systems [5]. These models could also be used to support periodic safety and security risk assessments and mandatory risk assessments when environments or procedures change, and/or when new communication processes or technologies are implemented. Also, policy makers could use these models to support the identification of mandatory regulations and standards with respect to communication for emergency prevention and management. In conclusion, these are promising developments and the incorporation of further psychological insights into crowd simulations will help enhance the realism of these models and the accuracy with which they can predict and prevent crowd disasters.

Acknowledgments. This research was undertaken as part of the EU HORIZON 2020 Project IMPACT (GA 653383) and Science without Borders – CNPq (scholarship reference: 233883/2014-2). We would like to thank our Consortium Partners and stakeholders for their input and the Brazilian Government.

References

1. Bosse, T., Duell, R., Memon, Z.A., Treur, J., van der Wal, C.N.: Agent-based modelling of emotion contagion in groups. Cog. Comp. J. **7**, 111–136 (2015)
2. Bosse, T., Hoogendoorn, M., Klein, M.C.A., Treur, J., van der Wal, C.N., van Wissen, A.: Modelling collective decision making in groups and crowds: integrating social contagion and interacting emotions, beliefs and intentions. Auton. Agnts. Mult. Agnt. Syst. J. **27**, 52–84 (2013)
3. Bradford City Stadium Fire. www.nist.gov/customcf/get_pdf.cfm?pub_id=100988&_ga=1.234008670.1199484209.1473418883
4. Challenger, R., Clegg, C.W., Robinson, M.A.: Understanding crowd behaviours. Practical Guidance and Lessons Identified, vol. 1. Cabinet Office, London (2010)
5. Clegg, C.W., Robinson, M.A., Davis, M.C., Bolton, L., Pieniazek, R., McKay, A.: Applying organizational psychology as a design science: a method for predicting malfunctions in socio-technical systems (PreMiSTS). Des. Sci. **3**, e6 (2017)
6. Damasio, A.: The Feeling of What Happens. Body and Emotion in the Making of Consciousness. Harcourt Brace, New York (1999)
7. Donald, I., Canter, D.: Intentionality and fatality during the King's cross underground fire. Eur. J. Soc. Psychol. **22**, 203–218 (1992)
8. Drury, J., Cocking, C., Reicher, S.: Everyone for themselves? A comparative study of crowd solidarity among emergency survivors. Br. J. Soc. Psychol. **48**(3), 487–506 (2009)
9. Duives, D.C., Daamen, W., Hoogendoorn, S.P.: State-of-the-art crowd motion simulation models. Transp. Res. Part C Emerg. Technol. **37**, 193–209 (2013)
10. Eagly, A.H., Crowley, M.: Gender and helping behavior: a meta-analytic review of the social psychological literature. Psychol. Bull. **100**(3), 283–308 (1986)
11. Fang, Z., Lo, S.M., Lu, J.A.: On the relationship between crowd density and movement velocity. Fire Saf. J. **38**(3), 271–283 (2003)
12. Formolo, D., van der Wal, C.N.: Simulating collective evacuations with social elements. In: Nguyen, N.T., Papadopoulos, G.A., Jędrzejowicz, P., Trawiński, B., Vossen, G. (eds.) ICCCI 2017, Part I. LNCS, vol. 10448, pp. 160–171. Springer, Cham (2017). https://doi.org/10.1007/978-3-319-67074-4_16
13. Galea, E., Deere, S., Filippidis, L.: The safeguard validation data set - sgvds1 a guide to the data and validation procedures, fire safety engineering group. University of Greenwich (2012)
14. Grosshandler, W.L., Bryner, N., Madrzykowski, D., Kuntz, K.: Draft report of the technical investigation of The Station nightclub fire. U.S. Department of Commerce Report (2005)
15. Hall, E.T.: A system for the notation of proxemic behavior. Am. Anthropol. **65**(5), 1003–1026 (1963)
16. Helbing, D., Buzna, L., Johansson, A., Werner, T.: Self-organized pedestrian crowd dynamics: experiments, simulations, and design solutions. Transp. Sci. **39**, 1–24 (2005)
17. Helbing, D., Johansson, A.: Pedestrian, crowd and evacuation dynamics. In: Meyers, R.A. (ed.) Extreme Environmental Events: Complexity in Forecasting and Early Warning, pp. 697–716. Springer, New York (2011). https://doi.org/10.1007/978-1-4419-7695-6_37
18. Hughes, H.P.N., Clegg, C.W., Robinson, M.A., Crowder, R.M.: Agent-based modelling and simulation: the potential contribution to organizational psychology. J. Occup. Organ. Psychol. **85**(3), 487–502 (2012)
19. Isobe, M., Helbing, D., Nagatani, T.: Experiment, theory, and simulation of the evacuation of a room without visibility. Phys. Rev. E **69**(6), 066132 (2004)

20. Kirchner, A., Schadschneider, A.: Simulation of evacuation processes using a bionics-inspired cellular automaton model for pedestrian dynamics. Physica A **312**, 260–276 (2002)
21. Kobes, M., Helsoot, I., De Vries, B., Post, J.G.: Building safety and human behaviour in fire: a literature review. Fire Saf. J. **45**(1), 1–11 (2010)
22. Liu, S., Yang, L., Fang, T., Li, J.: Evacuation from a classroom considering the occupant density around exits. Phys. A Stat. Mech. App. **388**(9), 1921–1928 (2009)
23. McConnell, N.C., Boyce, K.E., Shields, J., Galea, E.R., Day, R.C., Hulse, L.M.: The UK 9/11 evacuation study: analysis of survivors' recognition and response phase in WTC1. Fire Saf. J. **45**(1), 21–34 (2010)
24. Moussaïd, M., Perozo, N., Garnier, S., Helbing, D., Theraulaz, G.: The walking behaviour of pedestrian social groups and its impact on crowd dynamics. PLoS ONE **5**(4), e10047 (2010)
25. Netlogo. https://ccl.northwestern.edu/netlogo/
26. Owen, M., Galea, E.R., Lawrence, P.J.: The exodus evacuation model applied to building evacuation scenarios. J. Fire. Prot. Eng. **8**(2), 65–86 (1996)
27. Parisi, D.R., Dorso, C.O.: Morphological and dynamical aspects of the room evacuation process. Phys. A Stat. Mech. App. **385**(1), 343–355 (2007)
28. Parisi, D.R., Dorso, C.O.: Microscopic dynamics of pedestrian evacuation. Phys. A Stat. Mech. App. **354**, 606–618 (2005)
29. Population distribution EU. http://ec.europa.eu/eurostat/statistics-explained/index.php/Population_structure_and_ageing
30. Proulx, G., Fahy, R.F.: The time delay to start evacuation: review of five case studies. Fire Saf. Sci. **5**, 783–794 (1997)
31. Qiu, F., Hu, X.: Modeling group structures in pedestrian crowd simulation. Sim. Mod. Pract. Th. **18**(2), 190–205 (2010)
32. Rao, A.S., Georgeff, M.P.: BDI agents: from theory to practice. In: ICMAS, vol. 95, pp. 312–319 (1995)
33. Reininger, B.M., Raja, S.A., Carrosco, A.S., Chen, Z., Adams, B., McCormick, J., Rahbar, M.H.: Intention to comply with mandatory hurricane evacuation orders among persons living along a coastal area. Disaster Med. Publ. Health Preparedness **7**(1), 46–54 (2013)
34. Rizzolatti, G., Sinigaglia, C.: Mirrors in the Brain: How Our Minds Share Actions and Emotions. Oxford University Press, Oxford (2008)
35. Ronen, S., Shenkar, O.: Mapping world cultures: cluster formation, sources and implications. J. Int. Bus. Stud. **44**(9), 867–897 (2013)
36. Santos, G., Aguirre, B.E.: A critical review of emergency evacuation simulation models (2004)
37. Soto, C.J., John, O.P., Gosling, S.D., Potter, J.: Age differences in personality traits from 10 to 65: big five domains and facets in a large cross-sectional sample. J. Pers. Soc. Psychol. **100**, 330–348 (2011)
38. Still, G.K.: Introduction to Crowd Science. CRC Press, Boca Raton (2014)
39. Templeton, A., Drury, J., Philippides, A.: From mindless masses to small groups: conceptualizing collective behavior in crowd modeling (2015)
40. Treur, J.: Network-Oriented Modeling: Addressing Complexity of Cognitive, Affective and Social Interactions. Springer, Switzerland (2016)
41. Tsai, J., Bowring, E., Marsella, S., Tambe, M.: Empirical evaluation of computational emotional contagion models. In: Vilhjálmsson, H.H., Kopp, S., Marsella, S., Thórisson, K.R. (eds.) IVA 2011. LNCS, vol. 6895, pp. 384–397. Springer, Heidelberg (2011). https://doi.org/10.1007/978-3-642-23974-8_42

42. Varas, A., Cornejo, M.D., Mainemer, D., Toledo, B., Rogan, J., Munoz, V., Valdivia, J.A.: Cellular automaton model for evacuation process with obstacles. Phys. A Stat. Mech. App. **382**(2), 631–642 (2007)

43. van der Wal, C.N., Formolo, D., Bosse, T.: An agent-based evacuation model with social contagion mechanisms and cultural factors. In: Benferhat, S., Tabia, K., Ali, M. (eds.) IEA/AIE 2017, Part I. LNCS, vol. 10350, pp. 620–627. Springer, Cham (2017). https://doi.org/10.1007/978-3-319-60042-0_68

44. Wei-Guo, S., Yan-Fei, Y., Bing-Hong, W., Wei-Cheng, F.: Evacuation behaviors at exit in CA model with force essentials: a comparison with social force model. Phys. A Stat. Mech. App. **371**(2), 658–666 (2006)

45. Wikipedia. List of countries by English speaking population (2017). https://en.wikipedia.org/wiki/List_of_countries_by_English-speaking_population

46. Willis, A., Gjersoe, N., Havard, C., Kerridge, J., Kukla, R.: Human movement behaviour in urban spaces: Implications for the design and modelling of effective pedestrian environments. Environ. Plann. B Plann. Des. **31**(6), 805–828 (2004)

47. Yuan, W.F., Tan, K.H.: An evacuation model using cellular automata. Physica A **384**, 549–566 (2007)

48. Zheng, X., Zhong, T., Liu, M.: Modeling crowd evacuation of a building based on seven methodological approaches. Build. Environ. **44**(3), 437–445 (2009)

Group Approximation of Task Duration and Time Buffers in Scrum

Barbara Gładysz$^{(\boxtimes)}$ and Andrzej Pawlicki

Faculty of Computer Science and Management, Wroclaw University of Science and Technology, Smoluchowskiego 25, 50-372 Wroclaw, Poland
{barbara.gladysz,andrzej.pawlicki}@pwr.edu.pl

Abstract. Expansion of modern IT technologies, which took place last years, caused a significant increase in software projects. Those projects are quite often complex venture burdened with high risk. Nowadays, a large number of software projects is managed using Scrum framework. In Scrum, where people form self-organizing team, group decisions became an essential element of the project, which plays an important role to create time approximation or to manage potential risks. This paper focuses on group decisions, temporal aspects of estimation and risk management in Scrum project. In article we present conceptual model of extension Scrum framework by risk management process in aspect of project time estimation. Proposed model contains time buffers based on mixture probability distribution, which improve Scrum framework in terms of group estimation. We also depict case study which presents time approximation process which took place in one of a Scrum project.

Keywords: Group decision · Time approximation · Risk management · Scrum · Agile

1 Introduction

In the 20th century, software projects were managed using traditional waterfall approach based on i.e. PRINCE2 methodology or PMBOK Guide. Sponsors of the project demanded huge amount of detailed documentation and expected working product at the end of a project. At the beginning of 21st century, in 2001, seventeen software experts decided to discover better way of software development, which had to be an alternative approach to traditional waterfall project management. Software experts determined lightweight development method and through this work they created Manifesto for Agile Software Development called Agile Manifesto [1]. Software experts agreed that in project, based on self-organizing and cross-functional teams, people should work using an incremental, iterative approach, instead of in-depth planning. Teams should be also open to changing requirements over time and encourage constant feedback from the end users. Work in accordance with the mentioned values was described as four fundamental Agile Manifesto values [1]:

1. Individuals and interactions over processes and tools;
2. Working software over comprehensive documentation;

© Springer International Publishing AG 2017
N.T. Nguyen et al. (Eds.): TCCI XXVII, LNCS 10480, pp. 178–190, 2017.
https://doi.org/10.1007/978-3-319-70647-4_12

3. Customer collaboration over contract negotiation;
4. Responding to change over following a plan.

From the beginning, basic assumption for Agile methodologies was that they are based on empiricism, iteration programming and observation that client's requirements quite often are evaluating during the project. In Agile projects, product is delivered incrementally from the start of the project, instead of trying to deliver it all at once near the end.

1.1 Scrum

Nowadays, there are various Agile software development methods. According to survey placed in *2015 State of Scrum report* published by "Scrum Alliance" organization, Scrum is most popular Agile practice [2]. Nearly all respondents (95%) reported that Scrum is used as their organization's Agile approach. The three other most common are Kanban (43%), Lean (21%) and Extreme Programming (13%), respectively. 54% respondents declared that they use Scrum in combination with other practices, while 42% reported exclusive use of Scrum. In research, multiple answers were allowed.

According to Schwaber and Sutherland "Scrum is a framework within which people can address complex adaptive problems, while productively and creatively delivering products of the highest possible value" [3]. Schwaber and Sutherland formed the Scrum process in 1995 when they formalized Scrum in order to present it at the "Oopsla conference". Researchers inherited the name Scrum from the 1986 groundbreaking paper *The New Product Development Game* presented by Takeuchi and Nonaka [4]. Authors noticed that in product development, speed and flexibility are essential and referred to the game of rugby to stress the importance of teams and some analogies between a team sport like rugby and being successful in the game of new product development game. The researchers described under the rugby approach, the product development process emerges from the constant interaction of a hand-picked, multidisciplinary team whose members work together from the start till the end. Takeuchi and Nonaka learned, from interviews with organization members, that leading companies show six characteristics in managing their new product development processes:

- Built-in instability;
- Self-organizing project teams;
- Overlapping development phases;
- Multilearning;
- Subtle control;
- Organizational transfer of learning.

Researchers noticed that "these characteristics are like pieces of a jigsaw puzzle. Each element, by itself, does not bring about speed and flexibility. But taken as a whole, the characteristics can produce a powerful new set of dynamics that will make a difference" [4]. Authors realized, that complex products are faster achieved when teams work as small and self-organizing units of people which can create group estimation of project activity times and as a whole unit people may discuss the risks that may occur. Based on Takeuchi and Nonaka research Sutherland and Schwaber presented the rules of Scrum by defining values, roles, ceremonies and artifacts.

Regarding the roles *The Scrum Guide* defines that Scrum Team consists of three roles.

- Product Owner – customer representative who build and maintain a relationship with the stakeholders and clearly communicate the business requirements to Development Team. This person is also responsible for creating, managing and maintaining the Product Backlog.
- Development Team – cross-functional and self-organized group of people who deliver the product increment.
- Scrum Master – person who removes impediments to help Development Team become productive. The person who coach Development Team, Product Owner and organization to follow the rules of Scrum.

The Scrum Guide defines that there are three artifacts in Scrum.

- Product Backlog – created, managed and maintained by the Product Owner, list of product requirements, called Product Backlog Items (PBIs). The Product Owner orders the PBIs according the priority and business intention. PBIs should have the attributes of order, priority, description and estimation. PBIs which may be expressed as a User Stories which have the same meaning as requirements, this is just another term to describe functional specification of the product [5].
- Sprint Backlog – consists of PBIs selected for the Sprint and a plan for delivering these items. Usually, it consists of tasks, which are the breakdown of PBIs. The Sprint Backlog is the work that Development Team will do to turn selected PBIs into a done increment.
- Increment – sum of all the PBIs completed during a Sprint and the value of the increments of all previous Sprints.

What is important is the fact that all Scrum artifacts should be as clear and simple as it is possible to maximize transparency so that everybody has mutual understanding of the artifact.

The Scrum Guide defines the following events.

- Sprint – time-box of one month or less period of time during which Scrum Team is creating and delivering product increment. Sprint consists of the Sprint Planning, Daily Scrums, Product Backlog Refinement, Sprint Review, and Sprint Retrospective meetings.
- Sprint Planning – every Sprint starts with a Sprint Planning meeting. Development Team works with the Product Owner to plan the work for the upcoming Sprint. They collaborate to help Development Team select Product Backlog Items (requirements) from the Product Backlog for the upcoming Sprint.
- Daily Scrum – a 15-minute time-boxed event for the Development Team. It is an opportunity to synchronize and inspect the work, adapt plans for the next 24 h and report obstacles or impediments. Anyone can attend Daily Scrum meeting, but only Development Team can participate.

- Product Backlog Refinement – it is the act of adding detail, estimates, and order to items in the Product Backlog when PBIs are reviewed and revised. "Refinement usually consumes no more than 10% of the capacity of the Development Team" [3].
- Sprint Review – a meeting which takes place at the end of the Sprint. Development Team with the Product Owner and stakeholders work together to see what was done during the Sprint and to update the Product Backlog for the future work.
- Sprint Retrospective – a final meeting in the Sprint. Development Team, Product Owner and the Scrum Master work together to reflect the previous Sprint i.e. what went well during the Sprint, what didn't go so well and what can they do to improve it. Retrospective meeting provides an opportunity for continuous improvement.

"All events are time-boxed, such that every event has a maximum duration. Once a Sprint begins, its duration is fixed and cannot be shortened or lengthened" [3].

1.2   Risk Management in Scrum

Large expansion of IT technologies can be observed in recent years which caused a significant increase in software projects managed using Scrum framework. Singh and Saxena remarked that nowadays methodologies which are used in IT projects are changing from traditional waterfall model to Agile approach [6]. Scrum framework "employs an iterative, incremental approach to optimize predictability and control risk" [3] so *The Scrum Guide* does not define formal risk management process. Despite this, nowadays "lack of risk management in Scrum projects is identified by various authors" [7]. Tomanek and Juricek underlined typical risk management processes deficiency like i.e. risk management project time. Researchers noticed that there is a lack of formal risk management processes in Scrum and proposed conceptual Risk Management model based on Scrum and PRINCE2 methodology. The authors conducted a survey to find out the current risk management practices in Agile projects. As a final result of the paper, the researchers proposed to extend the conceptual framework, based on Scrum and PRINCE2, by risk management processes.

Reddaiah et al. noticed that project risks can occur in Scrum projects, but Scrum formally has no risk management process implemented [8]. The authors remarked that Scrum does not define practices which can identify the reason of risks that may appear and underlined that Scrum does not guide how to evaluate possible risk impact on the project and manage the effort which is needed to mitigate the risk. In the study, the authors suggest a new concept for risk management process by adding to Scrum framework a new team which can be helpful as the project is characterized with a high risk factor. A new team would be managed by Scrum Master and its main responsibilities would be risk identification and risk management. The new team would also be responsible for finding a solution how to project risk can be mitigated immediately.

The successful researchers who raise the risk management in Scrum subject are Uikey and Suman. The authors presented an authorial framework, called "Risk Based Scrum Method", which aims to incorporate risk management processes to improve the Scrum method [9]. The researchers underlined, that the proposed approach may increase project planning quality and performance schemes. Uikey and Suman recommend concept of the "RBSM" model which may potentially reduce the difficulty of a Scrum project.

Andrat and Jaswal presented another alternative approach for risk management in Scrum presented. In the paper [10], the authors noticed that Scrum is the most popular approach in comparison to other Agile methodologies and highlighted how risk management process is important in software projects. The researchers underlined that even in projects managed using Scrum there is a place for further improvements. The authors precisely explored Scrum framework and proposed their own model which may be an alternative approach for risk assessment in Scrum.

The purpose of this paper is to analyze temporal aspects of approximation and review risk management models proposed by various researchers for Scrum framework. Janczura and Kuchta proposed estimation approach based on fuzzy numbers [11]. Cahierre et al. analyzed time buffers as a tool for stability assurance of the project schedule and proposed to determine time buffers by experts [12]. Our model contains time buffers (lower and upper) based on mixture probability distribution, whereby progress of work may be reviewed and in case of delay number of people on the team or budget can be increased. In case of faster completion of work product quality can be increased or managers may assign people to the other projects. In the article, we propose a model for time estimation which may extend Scrum framework by risk management process in aspect of project time estimation. The proposed model may improve group approximation decision in Scrum project for time estimation for requirements placed in the Product Backlog. At the end of the paper we present the case study which describes practical adoption of proposed model in Scrum project.

2   Group Approximation of Time in Scrum

In Agile methodologies, it is assumed that the group decisions made by a team are better than the decisions of individuals. In Scrum projects, group decision, including approximation, is a very common practice. It is used when there is a need for a variety of information or broad expertise is required in various fields or there may appear a significant risk in a project. Group decision has its advantages and as well limitations. The advantage is that more people can be involved in decision-making, provide more information and more options to solve the problem. The disadvantage is that more people can participate in decision-making which may result in prolonged decision-making process, which can lead to lengthening the time required for analysis and decision making.

Scrum eliminates traditional technical role such as an architect, as technical decisions are made by collaborative team. In Scrum projects group estimations are commonly used by "cross-functional teams who have all competencies needed to accomplish the work without depending on others not part of the team" [3]. Having all the necessary competencies it is a prerequisite to achieve time estimation for PBIs by Development Team, to present work effort for Product Owner and project sponsors. Group decision making is also one of the key element of Sprint Review meeting agenda when the entire group: Development Team, Product Owner, Scrum Master and key stakeholders, collaborates on what to do in the next Sprint.

Time approximation is a difficult but a very important part of Scrum development process. Important because Scrum Team is using estimation during the whole project, not only in one iteration, and estimations may be used as a reference in further Sprints. All the entries, within the Product Backlog, have to be estimated to allow the Product Owner to prioritize the entries and to plan Release timeline.

2.1 Scrum Estimation Techniques

Scrum estimation techniques are collaborative and are designed to be fast. Whole Scrum Team is included in the process of estimating effort of PBIs, so that it is impossible to blame someone for an incorrect estimate. There are various estimation methods used in Scrum projects like i.e.: Planning Poker, T-Shirt Sizes, Relative Mass Valuation, Bucket System or Dot Voting [13].

2.2 Planning Poker

As an example of an approximation method we present Planning Poker, a game often played by the Scrum Team. Planning Poker as Scrum estimating technique helps to estimate PBIs complexity. Planning Poker support, in a secure way, to prepare accurate estimates for PBIs as a fine result before Sprint Planning meeting is held. Planning Poker is very simple and very effective at the same time. Most Scrum Teams hold a Planning Poker session shortly after an initial Product Backlog is written. This session is used to create initial estimates useful in scoping or sizing the Release, because PBIs will continue to be added throughout the whole Release. Most Scrum Teams find it helpful to conduct subsequent Scrum estimating and planning session during Sprint Product Backlog Refinement.

One of the basic rules of the Planning Poker is that each Development Team member takes place in the game to be sure that all members are able to provide their own estimation. Product Owner and Scrum Master are not allowed to vote for PBIs to avoid pressuring (intentionally or otherwise) the Development Team. The Scrum Master is responsible for facilitating the meeting and keeping it time-boxed. The Product Owner (usually as a moderator) presents each PBI for Development Team to be estimated, explains PBI in more detail way (if necessary) and answers the questions, to give an opportunity to discuss and clarify assumptions.

Each Development Team member gets an identical set of cards with values like: 0, 1, 2, 3, 5, 8, 13, 20, 40 and 100. Cards are used to set value of Story Point for each PBI. The "0" card means "this functionality requires just a few minutes of work" or "this story is already done". There are also two additional cards with coffee cup symbol (which means: "coffee break") and a question mark symbol (which means "I have no idea"). Question mark card should be rare. If this card is used too often, the Scrum Team needs to discuss the PBI more and try to achieve their better knowledge. When the estimation begins for each PBI Product Owner, as a moderator, reads the description and then each Development Team member call one card simultaneously by turning it over and presenting the estimated value, so that all participants can see one another's estimate. What is important, estimation for each PBI should be presented by

Development Team member at the same time, which can avoid the influence of the other participants.

If all estimators selected the same value, that becomes the estimate. But due to the fact that each member of the Development Team has commonly different experience, it is very unlikely that everyone will come up with the same estimation. So if the estimates are different (most common situation) people with highest and lowest estimates are allowed to explain their estimate, trying to avoid defending own estimation at all cost, but rather by trying to provide arguments for presented value. If Development Team cannot reach an agreement then various techniques may be used. Some of the teams use average value and some use median value. What is noteworthy is the fact that median value rejects the extreme values which, due to different Development Team members' expertise, can sometimes be best estimated value. Some of researchers say that average value is not the best estimation method, by saying that median value is better [14]. Our suggestion is to take into account the estimates of all Development Team members through the use of mixture probability distribution. Taking into account, all members of the team rating, allows to determine buffers for time approximation for Sprint and for whole Release as well.

To sum up, logic behind the Planning Poker is simple and there are many advantages of the game. Playing Poker is based on the wisdom of crowds, as it is possible to get the benefit of the Development Team's collective intelligence. Planning Poker leads to better estimates because it brings together a group of expert opinions. Experts form a cross-functional Development Team from all disciplines and they are better suited to the estimation task than anyone else. Planning Poker technique helps to avoid the influence of the other participants, that is why this technique forces people to think independently and tell their numbers simultaneously. What is important: the game is quick and dynamic which allows to estimate many PBIs in limited time. What is also important is the fact that Planning Poker game is some kind of escape from formal meetings and daily duties, so Planning Poker makes the Development Team's work more effective and enjoyably what is recommended by *The Scrum Guide*. In Planning Poker game everyone gets a chance to speak, there is a chance for group discussions and averaging individual estimates lead to better estimates.

3 Method for Task Duration and Time Buffers Estimation in Scrum

Malcolm et al. [15] assume that the time required to carry out a task has a beta distribution $Beta(a, b, \alpha, \beta)$ on the interval $[a, b]$ with density function:

$$f(t) = \frac{1}{B(\alpha, \beta)} \frac{(t - a)^{\alpha-1}(b - t)^{\beta-1}}{(b - a)^{\alpha+\beta-1}} \tag{1}$$

The expected value, variance of such a random variable are given as follows (for $\alpha > 1, \beta > 1$):

$$E(T) = \frac{\alpha b + \beta a}{\alpha + \beta} \tag{2}$$

$$D^2(T) = \frac{\alpha \beta (b-a)^2}{(\alpha + \beta)^2 (\alpha + \beta + 1)} \tag{3}$$

When $\alpha + \beta = 6$ the kurtosis of the beta distribution is equal to 3 and thus the distribution is somewhat similar in shape to the normal distribution, see [16] and [17]. In this case expected value and variance are given as follows:

$$E(T) = \frac{(3 \pm \sqrt{2})b + (3 \mp \sqrt{2})a}{6} \tag{4}$$

$$D^2(T) = \left(\frac{b-a}{6}\right)^2 \tag{5}$$

These pieces of information are used to construct mixture probability distribution $F(t)$, expected value $E(T)$ and variance $D^2(T)$ of task durations T, which is the core of our proposal:

$$F(t) = \sum_{i=1}^{n} w_i F_i(t) \tag{6}$$

$$E(T) = \sum_{i=1}^{n} w_i E(T_i) \tag{7}$$

$$D^2(T) = \sum_{i=1}^{n} w_i \left[(E(T_i) - E(T))^2 + D^2(T_i) \right] \tag{8}$$

where:

$F_i(t)$ – beta distribution on the interval $\left[t_i^F, t_i^L\right]$,

$E(T) = \frac{(3 + \sqrt{2})t_i^F + (3 - \sqrt{2})t_i^L}{6}$,

$D^2(T) = \left(\frac{t_i^F - t_i^L}{6}\right)^2$,

t_i^F (t_i^L) – task duration estimated by i-th Development Team member in the first round (last round); t_i^F – could be also the duration which is the most preferable by i-th Development Team member.

For $\alpha + \beta = 6$ beta distribution is somewhat similar in shape to the normal distribution, we can construct lower and upper buffers for task duration equal to:

$$3\sqrt{D^2(T)} \tag{9}$$

4   Case Study

Some Scrum Teams use Story Points for estimation and the others use hours, other do a mix [18], so values assigned to each PBI may be converted from Story Points into working days, working hours or other commonly agreed time unit. In the case study, time estimation was made using working hour unit and it was performed for Scrum web based project in Poland.

Project Scrum Team was engaged in the final work of an ongoing Release. Development Team was preparing delivered increment for deployment in production environment. At this moment project sponsors informed that the estimation had to be done for an upcoming Release which consisted of seven, two weeks long Sprints. Product Owner gathered list of six requirements from project sponsors and after placing them in the Product Backlog, asked Development Team, consisting of three members (Dev 1, Dev 2 and Dev 3), to provide time estimation for each PBI. Every two of six requirements were marked as critical, high, and medium in accordance to business needs (see Table 1).

Table 1. Development Team's time estimations for Product Backlog Items.

PBIs Product Backlog Items	Priority	Time estimation			
		First round			Last round
		Dev 1	Dev 2	Dev 3	(Dev 1, Dev 2, Dev 3)
Task no. 1	Critical	21	23	20	21
Task no. 2	Critical	2.5	2.5	2	2.5
Task no. 3	High	11	12	9	11
Task no. 4	High	5.5	7	6	6
Task no. 5	Medium	1	1	1	1
Task no. 6	Medium	28	31	27	29
Total					70.5

Time unit: working days.

Using Planning Poker game, each member of the Development Team presented his own time estimation for every PBI (see Table 1) using working day as the unit of time. Due to the fact that each Development Team member had different experience, estimations varied widely. After the discussion, Development Team members found consensus and provided common group time approximation.

As we can see, estimation time for the Release is 70.5 working days. Development Team used time buffers and set the risk margin at the level of –10% and +25% of estimated time, respectively lower and upper bounds were equal to 63 and 88 working days. In practice, when during the project it turns out that approximated time was underestimated usually Product Owner suggests the project sponsors to prolong the duration of Release or to remove some of the requirements, predominantly tasks with lowest priority.

In the presented case study, it occurred that experience which Development Team gained in previous Release was very useful and Development Team was able to finish the work earlier than it was estimated, within 60 working days (see Table 2) and also earlier than assumed lower bound 63 working days for the Release, thereby Development Team was able to start work earlier on requirements provided for the next Release.

Table 2. Expected times, lower and upper buffers and real times for Product Backlog Items.

PBIs Product Backlog Items	Priority	Expected time $E(T)$	Lower bound = $\left|E(T) - 3\sqrt{D^2(T)}\right|$	Upper bound = $\left|E(T) + 3\sqrt{D^2(T)}\right|$	Real time
Task no. 1	Critical	21	18	24	19
Task no. 2	Critical	2	2	3	2
Task no. 3	High	11	7	14	9
Task no. 4	High	6	5	7	5
Task no. 5	Medium	1	1	1	1
Task no. 6	Medium	29	25	33	24
Total		70	58	82	60

Time unit: working days.

Now let's apply proposed model for estimation time buffers in this case. Like it was mentioned, Development Team members had various experience, skills and knowledge of web application which was developed during the project. Dev 1 was a new member of the team with low technical skills, Dev 2 had some experience in the ongoing project and also in programming and Dev 3 was the most experienced person, both in the project and programming. Therefore, in formula for the expected time (Eq. (7)) and the buffers (Eq. (9)) we assigned the following weights: for Dev 1 – $w_1 = 0.15$, for Dev 2 – $w_2 = 0.35$, for Dev 3 – $w_3 = 0.5$. The results are presented in Table 2. The real time of Task no. 6 is smaller than its lower bound, the real times of other tasks are between their lower and upper bounds (see Table 2). In IT projects, which are realized in Scrum methodology, task duration is usually not independent. So in the worst case, lower (upper) bound of Release is a sum of lower (upper) bounds of all task durations in this Release. In our project, these lower and upper bounds are respectively 58 and 82 working days (see Table 2). As one can see, real time of 60 working days belongs to the interval of [58, 82] working days. So the estimated buffers, according to the proposed method, are more detailed and more compatible with the actual realization of the project than traditional buffers −10%, +25. Especially the upper buffer (+25%) is overestimated in our opinion.

Methods of estimating the time buffer are, above all, adapted to projects for which it is possible to derive a critical path or critical chain, e.g.: [19–22]. Mike Cohn [23] proposed adapting the approach of Reinertsen's method for estimating the problem of the planned duration and the time buffer for each task to project management using the

AGILE methodology. Reinertsen proposed that the time buffer for a task should be estimated on the basis of two parameters: the median (ME) and the 0.90 quantile ($Q_{0.9}$) of the distribution of the duration of a task. At the same time, he assumed that the difference between the estimators of these times is equal to two standard deviations of the duration of a task. The time buffers for individual tasks and the project as a whole can be calculated as follows:

The buffer for the i-th task:

$$buffer_i = \frac{Q_{0.90,i} - ME_i}{2}.$$ (10)

The buffer for the project as a whole:

$$buffer = 2\sqrt{\sum_{i=1}^{n} \left(\frac{Q_{0.90,i} - ME_i}{2}\right)^2}$$ (11)

Reinertsen assumed that the duration of the tasks are independent random variables. In addition, the planned duration of a task is taken to be the median of the duration of that task and the time buffers for individual tasks and the entire project, defined by Eqs. (10) and (11) respectively, serve as protection against the project not meeting the required deadline. Alternatively, the buffer for the entire project can be defined as the sum of the buffers for the individual tasks $buffer = \sum_{i=1}^{n} \frac{Q_{0.90,i} - ME_i}{2}$, see: [19, 23].

Table 3 presents the results of estimating the duration of the tasks using Reinertsen's method, taking into account time buffers based on estimates given by members of the Development Team of the analysed project. The actual durations of tasks are shorter than the lower estimate of the duration of those tasks based on the median (except for Task 5, which is a short task – assumed to last one day). Comparing the time characterization of the project obtained using Reinertsen's method with the method proposed in this article, it can be seen that the interval estimates of the duration of the tasks (lower bound, upper bound) are significantly narrower when Reinertsen's method is used (see Tables 2 and 3). Above all, this results from the fact that Reinertsen's method assumes that the durations of tasks are independent random variables and that their goal is to protect against a task, and the project as a whole, not meeting its deadline. In general, the actual durations of tasks belonged to the corresponding interval calculated according to the method proposed here. Only Task 6 was completed in a time shorter than the lower bound of the corresponding interval estimate of the duration.

When the buffer for the project as a whole is obtained by summing the buffers for the individual tasks, then we obtain analogous results for the characterization of the duration of the project as a whole.

The method proposed in this article assumes that the duration of a task is very likely to belong to the interval ($E(T) - 3\sqrt{D^2(T)}, E(T) + 3\sqrt{D^2(T)}$). In accordance with Chebyshev's inequality (the 3-sigma rule), the probability of the duration of a task belonging to this interval is at least $8/9 \approx 0.9$. We can obtain estimators of the expected duration $E(T)$ and variance $D^2(T)$ (Eqs. (7) and (8)) based on estimates given by individual members of the Development Team. Estimating the quantiles of a mixture of

Table 3. Estimates of the durations of tasks and buffers using Reinertsen's method

PBIs Product Backlog Items	Priority	Lower bound = $Me(T)$	Quantile $Q_{0.90}$	Upper bound = $Me(T) + buffer$	Real time
Task no. 1	Critical	21	23	22	19
Task no. 2	Critical	2.5	2.5	2.5	2
Task no. 3	High	11	12	11.5	9
Task no. 4	High	6	7	6.5	5
Task no. 5	Medium	1	1	1	1
Task no. 6	Medium	28	31	29.5	24
Total		70	69.5	23	73

distributions (see Eq. (6)) would require numerically intricate calculations. This is not in line with the assumptions of the SCRUM methodology, which assumes that all the techniques applied in project management should be simple. In addition, we do not assume that the durations of tasks in an IT project are independent random variables. The proposed method also takes into account the possibility of finishing a task more quickly than the planned duration (equal to the median). In practice, such a time buffer is applied to IT projects. For example, in the case study the limits on the time buffer were set to be −10% and +25% in comparison to the planned duration of a task.

Leach in [24] states that for a buffer to fulfil its role, its length should be at least 20% of the planned duration of the project. In our case study, the length of the buffer defined according to Reinertsen's method is only about 2% of the planned duration of the project. On the other hand, the difference between the estimates of the lower and upper bounds on the expected duration of the project derived using the method presented here is about 17% of the planned duration of the project.

5 Summary

In the paper, we focused on group decisions, time approximation buffers and risk management process. We proposed new model for time approximation risk margins which may be used in Scrum projects – nowadays most popular Agile approach. We presented the case study where we described how self-organizing Development Team, based on proposed model, can estimate time effort buffers for project requirements listed in PBIs form. Used margin, based on proposed model, proved to be correct in practice. The proposed model is simple and effective and can be used in projects managed using Scrum framework.

References

1. Manifesto for Agile Software Development (2001). http://agilemanifesto.org. Accessed Jan 2017
2. The 2015 State of Scrum Report. Scrum Alliance (2015)

3. Schwaber, K., Sutherland, J.: The Scrum Guide. The Definitive Guide to Scrum: The Rules of the Game. Scrum.Org and ScrumInc. (2016)
4. Takeuchi, H., Nonaka, I.: The new new product development game: Stop running the relay race and take up rugby. Harvard Bus. Rev. 137–146 (1986)
5. Canty, D.: Agile for Project Managers. Best Practices and Advances in Program Management Series. CRC Press Tylor and Francis Group (2015)
6. Singh, M., Saxena, R.: Risk management in agile model. IOSR J. Comput. Eng. **16**, 43–46 (2014)
7. Tomanek, M., Juricek, J.: Project risk management model based on PRINCE2 and scrum frameworks. Int. J. Softw. Eng. Appl. **6**, 81–88 (2015)
8. Reddaiah, B., Ravi, S., Movva, L.: Risk management board for effective risk management in scrum. Int. J. Comput. Appl. **65**, 16–23 (2013)
9. Uikey, N., Suman, U.: Risk based scrum method: a conceptual framework. In: Hoda, M.N. (Ed.) Proceedings of the 9th INDIACom; INDIACom-2015, IEEE Conference ID: 35071 2015 2nd International Conference on "Computing for Sustainable Global Development", 11th–13th March 2015, Bharati Vidyapeeth's Institute of Computer Applications and Management (BVICAM), New Delhi, pp. 4.120–4.125 (2015)
10. Andrat, H., Jaswal, S.: An alternative approach for risk assessment in scrum. In: Proceedings of 2015 International Conference on Computing and Network Communications (CoCoNet 2015), Trivandrum, India, 16–19 December 2015, pp. 535–539 (2015)
11. Janczura, M., Kuchta D.: Proactive and reactive scheduling in practice. Research Papers of Wrocław University of Economics, Wrocław, vol. 238, p. 38 (2011)
12. Cahierre, R., Kuchta, D., Ślusarczyk, A.: Application of Buffers in Project Management. Research Papers of University of Economics in Katowice **235**, 34–46 (2015). in Polish
13. Singh, V.: Agile Estimation Techniques. Scrum Alliance. https://www.scrumalliance.org/community/articles/2016/january/agile-estimation-techniques. Accessed Jan 2017
14. Cohn, M.: Don't Average During Planning Poker. Mountain Goat Software (2007). https://www.mountaingoatsoftware.com/blog/dont-average-during-planning-poker. Accessed Jan 2017
15. Malcolm, D.G., Rosenboom, J.H., Clark, C.E., Fazar, W.: Application of a technique for research and development program evaluation. Oper. Res. **7**, 646–669 (1959)
16. Gallagher, C.A.: A note on PERT assumptions. Manage. Sci. **33**, 1360 (1987)
17. Kamburowski, J.: New validations of PERT times. Int. J. Manag. Sci. **25**, 323–328 (1997)
18. Mitchell, I.: Should we or should we not estimate tasks? Scrum.org. (2012). https://www.scrum.org/Forums/aft/229. Accessed Jan 2017
19. Goldratt, E.M.: Critical Chain. North River Press, Massachusetts (1997)
20. Reinertsen, D.G.: Managing the Design Factory: a Product Developers's Toolkit. Free Press, New York (1997)
21. Fallah, M., Ashtiani, B., Aryanezhad, M.B.: Critical chain project scheduling: utilizing uncertainty for buffer sizing. Int. J. Res. Rev. Appl. Sci. **3**, 280–289 (2010)
22. Cohn, M.: Agile Estimation and Planning. Robert C. Martin Series. Pearson Education Incorporation, Upper Saddle River (2006)
23. Gładysz, B., Kuchta, D., Skorupko, D., Duchaczek, A.: Project risk time management: a proposed model and a case study in the construction industry. Procedia Comput. Sci. **64**, 24–31 (2015)
24. Leach, L.P.: Critical Chain Project Management. Artech House, Norwood (2000)

Extending Estimation of Distribution Algorithms with Agent-Based Computing Inspirations

Aleksander Byrski$^{(\boxtimes)}$, Marek Kisiel-Dorohinicki, and Norbert Tusiński

Department of Computer Science, AGH University of Science and Technology,
Al. Mickiewicza 30, 30-059 Krakow, Poland
{olekb,doroh}@agh.edu.pl, norbert.tusinski@gmail.com

Abstract. In the paper several extensions of a successful EDA-type algorithm, namely $COMMA_{op}$, inspired by the paradigm of agent-based computing (EMAS) are presented. The proposed algorithms leveraging notions connected with EMAS, such as reproduction and death, or even the population decomposition, turn out to be better than the original algorithm. The evidence for this is presented in the end of the paper, utilizing QAP problems by Éric Taillard as benchmarks.

Keywords: Estimation of distribution algorithm · COMMA · Evolutionary computing · Evolutionary multi-agent system

1 Introduction

Tackling difficult optimization problems, often described as *black-box* [1] ones, one has to carefully plan the search taking advantage of different aspects of well-known meta-heuristics (i.e. higher level, general heuristics—algorithms that provide "good-enough", maybe not optimal, solutions in reasonable time), in order to achieve success. In many difficult cases, hybridization of different metaheuristics can bring new quality into the problem solving, thus following well-known no free lunch theorem [2], it is always a good time for proposing new, but carefully designed general optimization algorithms [3].

Difficult search problems (such as Travelling Salesman Problem, or Quadratic Assignment Problem or many others) are usually tackled by different types of evolutionary algorithms [4]. Particularly efficient results have been obtained using Estimation of Distribution Algorithms [5]. As evolutionary algorithms are usually centered around the principle of recombination of solutions by means of selection and crossover, in EDAs the solutions are sampled from probabilistic distributions that model features of selected solutions, usually the most promising ones.

One of successful EDAs is $COMMA$ proposed by Olivier Regnier-Coudert and John McCall [6,7]. The COMpetitive Mutating Agents (COMMA) algorithm performs exploration and exploitation phases at in parallel by using a

© Springer International Publishing AG 2017
N.T. Nguyen et al. (Eds.): TCCI XXVII, LNCS 10480, pp. 191–207, 2017.
https://doi.org/10.1007/978-3-319-70647-4_13

population of agents and assigning different roles to each, in particular using mostly mutation for generating new individuals with adaptation of its range coming from the geometric inspirations. Moreover, it is quite easy to see, that EDAs and in particular COMMA are quite closely related to the agent-based computing systems, as they leverage some of their notions, as e.g. autonomy in undertaking the decisions about individual mutations. Thus these algorithms can be treated as a good starting points for introducing hybrid computing methods.

In this paper these inspirations are drawn from an efficient, agent-based search and optimization system [8], that was proposed by Krzysztof Cetnarowicz in 1996 and extended many times since then: Evolutionary Multi-Agent System (EMAS). This system is composed of agents—pseudo-intelligent, autonomous objects [9], which are able to make decisions by themselves, based on an interaction with other agents and with environment. As main task is decomposed into sub-tasks, each of which is entrusted to an agent, EMAS is an effective implementation of distributed problem solving. To this day, EMAS proved to be much more efficient than classic evolutionary algorithm and was applied successfully to different problems—classic benchmarks [10], inverse problems [11] and other optimization tasks [12,13].

Thus several hybridizations of the $COMMA_{op}$ algorithm are presented and evaluated in this paper, using selected instances of Quadratic Assignment Problem. In the next section the general idea of EDAs is presented, and followed by the description of $COMMA$ and $COMMA_{op}$. Later the basic notions of evolutionary multi-agent computing are discussed and the proposed extensions for COMMA, inspired by EMAS are described. Finally the experimental results are presented and the paper is concluded.

2   Estimation of Distribution Algorithms

Estimation of Distribution Algorithms are universal metaheuristics stemming from Evolutionary Algorithms. As EAs, they use stochastic sampling of the solution space in order to produce new individuals (often called agents), however in the case of EDAs, the probability distribution is usually derived from the information gathered in the current population, and accordingly adapted [14]. General strategy realized in EDAs can be described as follows: where: pop_t stands for a population observed at an iteration t pos is the solution space, $pos_j \in pos$ is one solution (an agent's genotype when compared to EA), Step 3 consists in employing a predefined selection strategy (similar to the ones used in EAs), Step 4 consists in estimating a new probability distribution, based on the information gathered in the population of agents, finally in Step 5 a new population is sampled using the probability distribution determined in Step 4.

Thus EDAs perceive the optimization process as a series of incremental updates of a certain probabilistic model starting from a model of uniform distribution and finishing with a model generating solely global extrema (or their approximation). In an ideal case, the quality of the generated solutions will grow in time and after certain (hopefully reasonable) number of iterations, the algorithm will generate a global optimum (or its accurate enough approximation).

Algorithm 1. Pseudo-code of EDA

1: $pop \leftarrow random\_generation\_of\_agents()$
2: **while** not stop_condition() **do**
3:     $pop_t^{Se} \leftarrow select\_agents(pop_{t-1})$
4:     $P_t(pos) \leftarrow P(pos_j | pop_t^{Se})$
5:     $pop_t \leftarrow P_t(pos)$

Different EDAs consist in different versions of the above mentioned steps, however the general idea remains the same—the strategy of generating new agents is iteratively adapted in order to increase the quality of the new solutions [15].

3 $COMMA$ Algorithm

One of interesting EDA-type algorithms is COMMA (COMpeting Mutating Agents) devised by Olivier Regnier-Coudert and John McCall [7]. In this algorithm, the sampling distribution is constructed considering the population of agents sorted according to their fitness. In this algorithm, for each position pos_j in the population pop sorted in descending order for maximization, a mutation distance d_j is set such that for two agents at positions e and f, such as $d_e \leq d_f$ if $e < f$. As it may become beneficial to allow low-quality solutions to be accepted, a probability p_j is also set for each pos_j. Each agent a_i is initially assigned a random solution s_i. The population is then sorted by fitness. At each generation, each agent mutates s_i using the distance $dist_i \in [1, d_r]$ defined according to its position r in the population. This step is equivalent to sampling from a distribution centered around s_i whose variance depends on r. If the mutated solution s_{new} has a better fitness than s_i, a_i replaces s_i with s_{new}. If s_{new} has a poorer fitness than s_i, s_{new} only replaces s_i with certain probability p_r. The pseudocode of the original $COMMA$ algorithm is as follows. The sampling range is inversely proportional to the fitness, thus the agents with high fitness values (in relation to other members of the population) are mutated with lower range than the individuals with lower fitness values. Thus the whole algorithm exploits around "good" solutions, at the same time starting exploration around "worse" ones. As the authors report, the above version of COMMA performed well when learning Bayesian network structures [6], however in order to enhance the proposed algorithm, they have significantly improved it by introducing multiple variation operators, proposing $COMMA_{op}$ algorithm.

4 $COMMA_{op}$ Algorithm

$COMMA_{op}$ starts by random generation of a population pop. Moreover, the level of alteration $\rho_{j,k}$ associated with every possible pair $\{\phi_j, d_k\}$ of operator and mutation distance is calculated and used to initialize the operator selection scale α ($k \in \lfloor 1, n \rfloor$ and $j \in [1, n_{op}]$, where n_{op} is the number of single operators included in $COMMA_{op}$. This means that any value k can be considered

Algorithm 2. Pseudo-code of $COMMA$ [7]

1: Initialize pop of σ agents with random solutions, distance vector d of size σ and probability vector p of size σ
2: **while** not stop_condition() **do**
3:     sort pop by fitness
4:     **for** each agent $a_i, i \in [0, \sigma - 1]$ **do**
5:         get position r of a_i in pop
6:         Sample new solution s_{new} with fitness fit_{new} by mutating s_i with distance $dist_i$ selected with uniform probability from $[1, d_r]$
7:         **if** $fit_{new} > fit_i$ **then**
8:             $s_i \leftarrow s_{new}$
9:         **else**
10:            $s_i \leftarrow s_{new}$ with probability p_r

---

for the mutation distance d_k for a chosen operator throughout the search. α is ordered with respect to $\rho_{j,k}$. For non-deterministic operator, whose associated $\rho j, k$ takes its value within a range, the mean $\rho_{j,k}$ is used to order α. The maximum number of generations $maxGen$ is also calculated at this stage from the maximum number of fitness evaluations and the population size σ. Finally, the $fixedDistance$ parameter needs to be set. It defines for a given mutation distance, whether a mutation strictly uses the distance or whether it can also use any value lower than the distance. This is another approach implemented in order to reduce the effect of some operators with high associated level of alteration. Using a fixed distance can be geometrically interpreted as sampling solutions only from the edge of a distribution, rather than from the whole distribution when $fixedDistance$ is set to false. Following this initialization step and at each generation, the population of agents is ordered by fitness and the rate is updated, that is the rate is decreased at each generation. Each agent then needs to sample a new solution. This is done according to its position r in the population. A pair of operator and mutation distance is picked from α with respect to r, but also considering some variation represented by the lower and upper bounds r^- and r^+, with $r^-, r^+ \in [0, |\alpha| - 1]$. Introducing such variation ensures that all $\{\phi_j, d_k\}$ pairs can be used to generate solutions during the search. $d(r')$ and $\phi(r')$ stand for the mutation distance and the operator at the r'-th position in α. The pseudocode of the $COMMA_{op}$ algorithm is as follows [7]: This algorithm extensively described above, became the starting point for introducing enhancements inspired by the evolutionary multi-agent computing experiences encountered by the authors during their work up-to-date.

5   Agent-Based Evolutionary Computing

Evolutionary algorithms and similar population-based metaheuristics [3,4] are very common means of solving difficult, often black-box [1] problems. Classic evolutionary algorithms (see Fig. 1) simplify several mechanisms well-known from

Algorithm 3. Pseudo-code of $COMMA_{op}$ [7]

1: Initialize pop of σ agents with random solutions
2: Initialize operator selection scale α with pairs $\{\phi_j, d_k\}$ for all selected operators ϕ_j and mutation distances d_k
3: Order α by $\rho_{\{j, k\}}$
4: Initialize $gen \leftarrow 0$ and maxGen
5: Initialize $fixedDistance$
6: **repeat**
7:     Sort pop by fitness in descending order
8:     $rate \leftarrow 1 - \frac{gen}{maxGen+1}$
9:     **for** each agent $a_i, i \in [0, \sigma - 1]$ **do**
10:         $r \leftarrow$ position of a_i in pop
11:         $r^- \leftarrow \frac{r \cdot |\alpha|}{\sigma} - rate \cdot |\alpha|$
12:         **if** $r^- < 0$ **then** $r^- \leftarrow 0$
13:         $r^+ \leftarrow \frac{r \cdot |\alpha|}{\sigma} + rate \cdot |\alpha|$
14:         **if** $r^+ > |\alpha| - 1$ **then** $r^- \leftarrow |\alpha| - 1$
15:         $r' \leftarrow random(r^-, r^+)$
16:         **if** $fixedDistance$ **then**
17:             Sample new solution s_{new} with fitness fit_{new} by mutating s_i with operator $\phi(r')$ and distance $d(r')$
18:         **else**
19:             Sample new solution s_{new} with fitness fit_{new} by mutating s_i with operator $\phi(r')$ and distance $dist_i$ selected with uniform probability from $(1, d(r'))$
20:         **if** $fit_{new} > fit_i$ **then** $s_i \leftarrow s_{new}$
21:     gen++
22: **until** $stopping_condition()$

---

the biological evolution, e.g. full synchronization of the reproduction while in biological systems no such phenomenon occurs. Another example would be lack of global knowledge in actual biological systems, while in evolutionary algorithms a single entity manages the whole population and synchronizes the variation operators etc. In order to try to leverage more natural phenomena in computing, EMAS was proposed (first in [16], then extended in [12,17] and many other papers).

EMAS utilizes the totally distributed (uncontrolled globally) phenomena of death and reproduction for modelling the processes of selection and inheritance (Fig. 2). Agent in EMAS carries the genotype (each agent has its own) and tries to survive competing with its neighbors (by executing the evaluation action) and proliferate (by reproduction leveraging recombination and mutation). Moreover, an agent may possess some knowledge acquired during its life, which is not inherited, yet controls the actions of the agent (e.g. sexual preferences [13]). Usually EMAS is implemented in a multi-deme model, similar to parallel evolutionary algorithm (cf. [18]).

One of the most important aspects of EMAS is distributed mechanism of selection, based on existing of non-renewable resource that is exchanged between

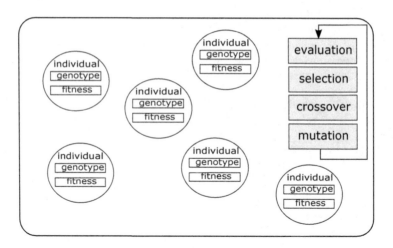

Fig. 1. Schematic presentation of Evolutionary Algorithm

the agents (in the process of evaluation—i.e. the better agent takes a certain part of the resource from the worse one). Later, the amount of the resource owned by a particular agent, becomes a condition for performing the action of reproduction (when the resource level is high) or death (when the resource level falls down). Moreover, the offspring overtakes certain part of the parents' energy during the reproduction.

EMAS prove to be an universal optimization tool (moreover, formal proofs of this feature have been constructed for EMAS and other computing systems, see, e.g. [19–21]), particularly with less demand on the number of the fitness function calls than its classic competitors, like PEA.

6   Proposed Extensions of COMMA

EMAS can be perceived as a system belonging to socio-cognitive class, because of clear inspirations taken from the area of social systems (usually all the agent systems can be counted to this class), and also cognitive ones (as the agents perceive the other agents, interact with them, can learn from them or get inspired by them). As elements of socio-cognitive related research have already proven to be effective in other computing related applications (namely discrete optimization using PSO [22] and ACO [23]), the authors have decided to try to enhance the EDA algorithm (as this class is very successful in solving discrete problems) with selected socio-cognitive mechanism, striving towards the full hybridization of EMAS and EDA. This approach seems to be right especially because individual adaptation of the mutation of each of agents makes the $COMMA$ algorithms sharing similar paradigm like EMAS—as this can be perceived as certain autonomy of the agents processed in $COMMA_{op}$. Therefore full hybridization of $COMMA_{op}$ and EMAS seems to be a promising idea.

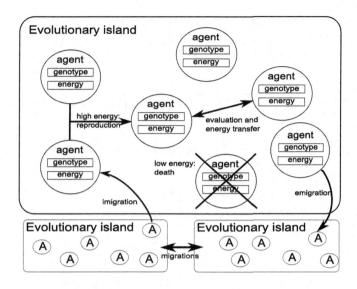

Fig. 2. Schematic presentation of EMAS

In order to clearly and precisely describe the modifications of the original algorithm, the pseudocode of the $COMMA_op$ is simplified by aggregating steps visible in Algorithm 4:

Algorithm 4. Simplified pseudo-code of $COMMA_{op}$ [7]

1: Initialization of agents and other parameters
2: **repeat**
3:     Sort population of agents
4:     Calculate mutation rates
5:     Sampling of new solutions by adaptive mutation
6: **until** $stopping_condition()$

6.1 $COMMA_op$ with Population Decomposition

Quite a natural evolution of $COMMA_{op}$ that may be considered following idea of EMAS and generally Parallel Evolutionary Algorithms concept [18] is the decomposition of the population that usually brings new quality with regards to the diversity of the search. Therefore the notion of evolutionary islands is introduced into the original algorithm, along with a simple migration strategy. Thus the mutation ranges are computed inside each of the islands. The modified algorithm can be described as follows (note that in this and the subsequent pseudocodes the changes introduced w.r.t. the previous version of the algorithm are displayed in bold): It is easy to see that in the beginning all the populations of agents are initialized on each of the islands. Then the migration is realized (with small probability) and the following step of the original $COMMA_{op}$ algorithm (cf. Algorithm 4) is realized on each of the islands subsequently.

Algorithm 5. Pseudo-code of $COMMA_{op}$ with population decomposition

---

1: Initialization of agents and other parameters
2: **repeat**
3:     **Migrate agents among the islands with low probability**
4:     Sort population of agents
5:     Calculate mutation rates
6:     Sampling of new solutions by adaptive mutation
7: **until** $stopping_condition()$

6.2 $COMMA_{op}$ with Cloning and Death of Agents

Another natural extension of $COMMA_{op}$ inspired by EMAS is introduction of the resource-based selection-like mechanism that reminds of the distributed selection in EMAS. The notion of energy is introduced and this value is computed for all of the agents, an action of energy exchange (similar to the one in EMAS) is used—the better agent takes a part of the energy from the worse agent— finally the notion of cloning of the agents is used, for those who exceed a certain amount of energy, and the notion of death—for those which energy falls down below a certain level. The pseudocode of this algorithm is as follows: The steps concerning cloning and mutation are realized on each of the island, along with the exchange of the energy between the agents.

Algorithm 6. Pseudo-code of $COMMA_{op}$ with cloning and death

---

1: Initialization of agents and other parameters
2: **repeat**
3:     Migrate agents among the islands with low probability
4:     **If agent's energy is higher than certain level: clone the agent in the population**
5:     **If agent's energy is lower than certain level: remove agent from the population**
6:     **Choose two agents and based on their fitness: exchange part of their energy**
7:     Sort population of agents
8:     Calculate mutation rates
9:     Sampling of new solutions by adaptive mutation
10: **until** $stopping_condition()$

6.3 $COMMA_op$ with Crossover

The final extension of $COMMA_op$ inspired by EMAS is introduction of the crossover with mutation (besides the original EDA-style mutation), instead of the cloning process: Thus the full hybrid of original $COMMA_{op}$ with EMAS-related notions is attained. Now one should turn to checking of the point of the whole endeavor.

Algorithm 7. Pseudo-code of $COMMA_{op}$ with crossover

1: Initialization of agents and other parameters
2: **repeat**
3:     Migrate agents among the islands with low probability
4:     **If agent's energy is higher than certain level: crossover of the agent**
       with another one on this island and mutate the offspring
5:     If agent's energy is lower than certain level: remove agent from the population
6:     Choose two agents and based on their fitness: exchange part of their energy
7:     Sort population of agents
8:     Calculate mutation rates
9:     Sampling of new solutions by adaptive mutation
10: **until** $stopping_condition()$

---

7   Experimental Results

This paper tackles Quadratic Assignment Problem (QAP) as a benchmark for testing the efficacy of the proposed COMMA extensions. The QAP is a combinatorial optimization problem stated for the first time by Koopmans and Beckman1 in 1957 [24]. It can be described as follows: Given two $n \times n$ matrices $A = (a_{ij})$ and $B = (b_{ij})$, find a permutation minimising:

$$min_{\pi \in \Pi(n)} f(\pi) = \sum_{i=1}^{n} \sum_{j=1}^{n} a_{ij} \cdot b_{\pi_i \pi_j} \tag{1}$$

where $\Pi(n)$ is the set of permutations of n elements. Shani and Gonzalez have shown that the problem is NP-hard and that there is no ϵ-approximation algorithm for the QAP unless P = NP [25]. It is to note that QAP instances of size larger than 20 are considered intractable and can be solved solely using heuristic approaches [26].

Three instances were taken from the QAP problems repository by Éric Taillard[1], namely `taillard15b`, `taillard20a` and `taillard20b`. Each of the experiments was repeated 10 times and average with standard deviation was shown along with minimum and maximum values of all repetitions.

---

[1] http://mistic.heig-vd.ch/taillard/problemes.dir/qap.dir/qap.html.

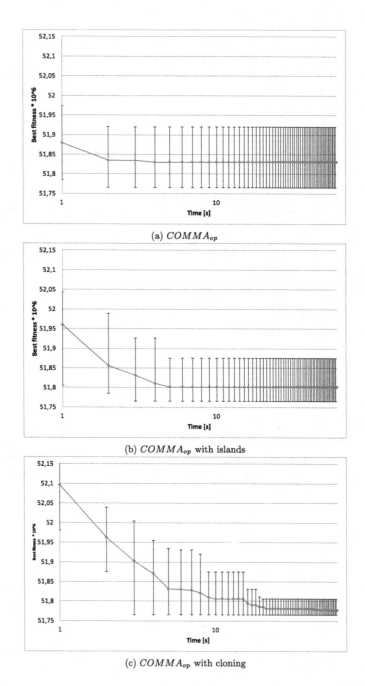

(a) $COMMA_{op}$

(b) $COMMA_{op}$ with islands

(c) $COMMA_{op}$ with cloning

Fig. 3. Fitness dependent on time for all tested algorithms for the problem taillard15b

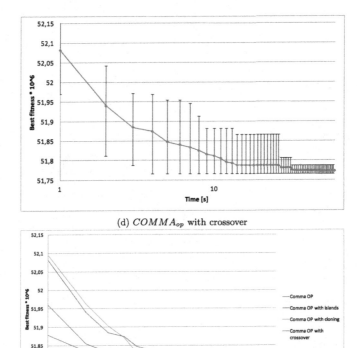

(d) $COMMA_{op}$ with crossover

(e) Comparison of all algorithms for taillard15b

Fig. 3. (*continued*)

The algorithms tested were:

- Original $COMMA_{OP}$,
- $COMMA_{OP}$ with ranking on islands,
- $COMMA_{OP}$ with cloning and dying of the agents,
- $COMMA_{OP}$ with crossover.

All of these algorithms were run for 60 s, the number of the agents equaled 50 and the number of islands (besides $COMMA_{OP}$) was 5. The graphs presented in this paper have minimum and maximum values of all of the runs pointed out. Moreover on all of the x-axes logarithmic scale was applied.

In Fig. 3 the results of optimization (fitness depending on time) for all the tested algorithms applied for the problem taillard15b are presented. It is easy to see that all the modification of the original algorithm surpassed the $COMMA_{op}$ and produced better optimization result. The best two results were produced using the modifications with cloning and crossover. Moreover, the dispersion of the results is significantly lower for all the modifications of the original

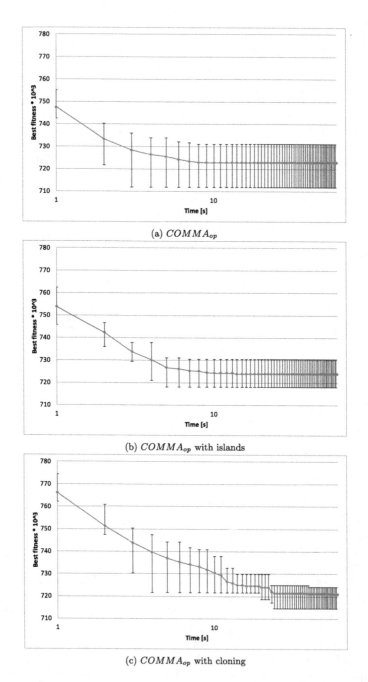

(a) $COMMA_{op}$

(b) $COMMA_{op}$ with islands

(c) $COMMA_{op}$ with cloning

Fig. 4. Fitness dependent on time for all tested algorithms for the problem taillard20a

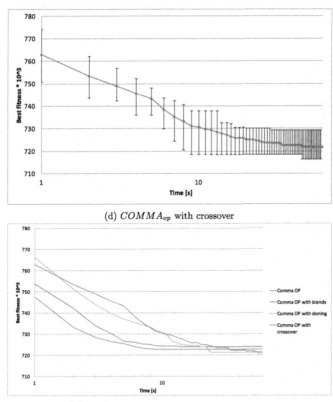

(d) $COMMA_{op}$ with crossover

(e) Comparison of all algorithms for taillard20a

Fig. 4. (*continued*)

$COMMA_{op}$, therefore it seems that the search using the new versions of this algorithm is more focused, and the results are more repeatable.

Quite a similar results can be observed for the second tested problem, namely taillard20a (see Fig. 4). Although the final outcome is not so clearly visible, still the modifications outrun the original algorithm. Again the cloning algorithm and the crossover are the best, in this case however the cloning one attains the result with the best fitness overall. The dispersion is still similar as in the previous experiment.

Finally the problem taillard20b is considered. In this case the dispersion of the original algorithm is lower, even lower than the dispersion of the algorithm with population decomposition (see Fig. 5). Regarding the results, this time all the outcomes are quite similar, however again the algorithm with cloning prevails in the end, although it can be fully verified after checking the results shown in Table 1, where the best overall results were emphasized with bold typeface.

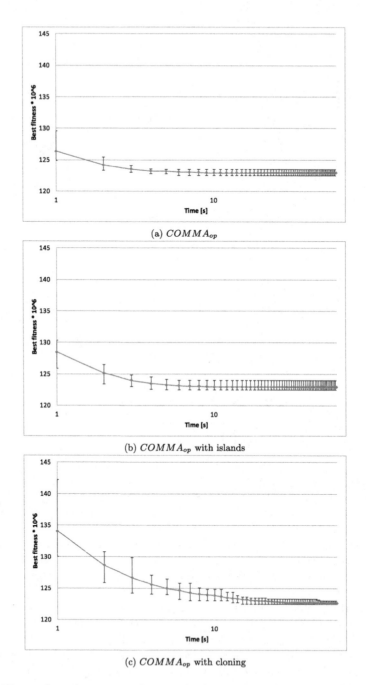

(a) $COMMA_{op}$

(b) $COMMA_{op}$ with islands

(c) $COMMA_{op}$ with cloning

Fig. 5. Fitness dependent on time for all tested algorithms for the problem taillard20b

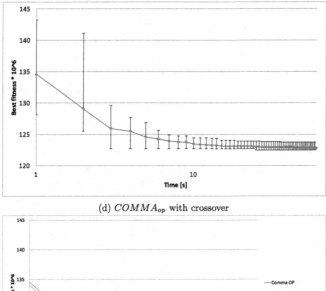

(d) $COMMA_{op}$ with crossover

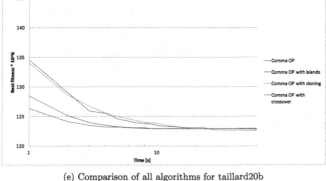

(e) Comparison of all algorithms for taillard20b

Fig. 5. (*continued*)

Table 1. Summary of final results attained by the tested algorithms

Algorithm	tai15b	tai20a	tai20b
$COMMA_{OP}$	51830404.4	722798.6	122927430.5
$COMMA_{OP}$ with ranking	51801686.3	723907.6	122953063.8
$COMMA_{OP}$ with cloning and dying	51777406.6	**720977.4**	**122616065.5**
$COMMA_{OP}$ with dual mutation	**51773297.2**	721495.4	122821966.9

8   Conclusion

In this work the novel modification of one of quite successful EDA-class algorithm, namely $COMMA_{op}$ by Olivier Regnier-Coudert and John McCall were presented. These modifications were based upon the fact of the closeness of the original algorithm (and EDAs in general) to the notions of agency—in particular autonomy in adaptation of mutation range per particular individual. Thus based on the authors' expertise in the field of evolutionary multi-agent computing, modifications based on introduction of population decomposition, cloning, crossover and death mechanisms were sketched out. These modifications were tested against the original algorithm using three QAP benchmark instances proposed by Éric Taillard and freely available on his website.

The obtained results show that such modifications can bring new quality into EDA research (in this case visible improvement of the results produced by $COMMA_{op}$ was pointed out). Moreover, the search conducted by the modified $COMMA_{op}$ algorithms tends to be more focused than the original algorithm.

In future more extensive testing of the introduced algorithms is planning, along with extension to other discrete and perhaps also continuous problems. Moreover, as in the case of the hybrid algorithm presented here, certain EMAS mechanisms were introduced into $COMMA_{op}$, a completely opposite approach is also possible and will be conducted: introducing EDA mechanisms into EMAS (e.g. by designing of a dedicated mutation strategy that will leverage the notions used in EDA, in other words it should individually adapt the mutation range based on e.g. agents encountered during the life of the one producing offspring in the course of the reproduction action).

Acknowledgment. This research was supported by AGH University of Science and Technology Statutory Fund.

References

1. Droste, S., Jansen, T., Wegener, I.: Upper and lower bounds for randomized search heuristics in black-box optimization. Theor. Comput. Syst. **39**, 525–544 (2006)
2. Wolpert, D., Macready, W.: No free lunch theorems for search. Technical report SFI-TR-02-010, Santa Fe Institute (1995)
3. Talbi, E.G.: Metaheuristics: From Design to Implementation. Wiley, Chichester (2009)
4. Michalewicz, Z.: Genetic Algorithms Plus Data Structures Equals Evolution Programs. Springer, New York (1994)
5. Larranaga, P., Lozano, J.: Estimation of Distribution Algorithms: A New Tool for Evolutionary Computation. Springer, US (2002)
6. Regnier-Coudert, O., McCall, J.: Competing mutating agents for Bayesian network structure learning. In: Coello, C.A.C., Cutello, V., Deb, K., Forrest, S., Nicosia, G., Pavone, M. (eds.) Parallel Problem Solving from Nature - PPSN XII, PPSN 2012. LNCS, vol. 7491, pp. 216–225. Springer, Heidelberg (2012). https://doi.org/10.1007/978-3-642-32937-1_22
7. Regnier-Coudert, O., McCall, J., Ayodele, M.: Geometric-based sampling for permutation optimization. In: Proceedings of the 15th Annual Conference on Genetic and Evolutionary Computation, GECCO 2013, pp. 399–406. ACM, New York (2013)
8. Byrski, A.: Agent-Based Metaheuristics in Search and Optimisation. AGH University of Science and Technology Press, Kraków (2013)
9. Kisiel-Dorohinicki, M., Dobrowolski, G., Nawarecki, E.: Agent populations as computational intelligence. In: Rutkowski, L., Kacprzyk, J. (eds.) Neural Networks and Soft Computing. Advances in Soft Computing, vol. 19, pp. 608–613. Physica, Heidelberg (2003). https://doi.org/10.1007/978-3-7908-1902-1_93

10. Byrski, A.: Tuning of agent-based computing. Comput. Sci. **14**(3), 491 (2013)
11. Wróbel, K., Torba, P., Paszyński, M., Byrski, A.: Evolutionary multi-agent computing in inverse problems. Comput. Sci. **14**(3), 367 (2013)
12. Dreżewski, R., Siwik, L.: Multi-objective optimization technique based on co-evolutionary interactions in multi-agent system. In: Giacobini, M. (ed.) EvoWorkshops 2007. LNCS, vol. 4448, pp. 179–188. Springer, Heidelberg (2007). https:// doi.org/10.1007/978-3-540-71805-5_20
13. Dreżewski, R., Siwik, L.: Co-evolutionary multi-agent system for portfolio optimization. In: Brabazon, A., O'Neill, M. (eds.) Natural Computing in Computational Finance. Studies in Computational Intelligence, vol. 100, pp. 271–299. Springer, Heidelberg (2008). https://doi.org/10.1007/978-3-540-77477-8_15
14. Ceberio, J., Irurozki, E., Mendiburu, A., Lozano, J.A.: A review on estimation of distribution algorithms in permutation-based combinatorial optimization problems. Prog. Artif. Intell. **1**(1), 103–117 (2012)
15. Pelikan, M., Hauschild, M., Lobo, F.: Introduction to estimation of distribution algorithms. Technical report 2012003, Missouri Estimation of Distribution Algorithms Laboratory (2012)
16. Cetnarowicz, K., Kisiel-Dorohinicki, M., Nawarecki, E.: The application of evolution process in multi-agent world (MAW) to the prediction system. In: Tokoro, M. (ed.) Proceedings of the 2nd International Conference on Multi-agent Systems (ICMAS 1996). AAAI Press (1996)
17. Byrski, A., Korczynski, W., Kisiel-Dorohinicki, M.: Memetic multi-agent computing in difficult continuous optimisation. In: KES-AMSTA, pp. 181–190 (2013)
18. Cantú-Paz, E.: A summary of research on parallel genetic algorithms. IlliGAL Report No. 95007. University of Illinois (1995)
19. Byrski, A., Schaefer, R., Smołka, M.: Asymptotic guarantee of success for multi-agent memetic systems. Bull. Pol. Acad. Sci.-Tech. Sci. **61**(1), 257–278 (2013)
20. Byrski, A., Schaefer, R.: Formal model for agent-based asynchronous evolutionary computation. In: 2009 IEEE Congress on Evolutionary Computation, pp. 78–85, May 2009
21. Schaefer, R., Byrski, A., Smolka, M.: The island model as a Markov dynamic system. Int. J. Appl. Math. Comput. Sci. **22**(4), 971–984 (2012)
22. Bugajski, I., Listkiewicz, P., Byrski, A., Kisiel-Dorohinicki, M., Korczynski, W., Lenaerts, T., Samson, D., Indurkhya, B., Nowé, A.: Enhancing particle swarm optimization with socio-cognitive inspirations. Procedia Comput. Sci. **80**, 804–813 (2016). International Conference on Computational Science 2016, ICCS 2016, 6–8 June 2016, San Diego, California, USA
23. Byrski, A., Świderska, E., Łasisz, J., Kisiel-Dorohinicki, M., Lenaerts, T., Samson, D., Indurkhya, B., Nowé, A.: Socio-cognitively inspired ant colony optimization. J. Comput. Sci. **21**(Suppl. C), 397–406 (2017). https://doi.org/10.1016/j.jocs.2016. 10.010
24. Koopmans, T., Beckmann, M.: Assignment problems and the location of economics activities. Econometrica **25**, 53–76 (1957)
25. Shani, S., Gonzalez, T.: P-complete approximation problems. J. ACM **23**, 555–565 (1976)
26. Gambardella, L.M., Taillard, É.D., Dorigo, M.: Ant colonies for the quadratic assignment problem. J. Oper. Res. Soc. **50**(2), 167–176 (1999)

Author Index

Printed in the United States
By Bookmasters